Context-Aware Pervasive Systems

T0133143

Context-Aware Pervasive Systems

Architectures for a New Breed of Applications

SENG LOKE

CRC Press
Taylor & Francis Group
Boca Raton London New York

CRC Press is an imprint of the
Taylor & Francis Group, an **informa** business

AN AUERBACH BOOK

CRC Press
Taylor & Francis Group
6000 Broken Sound Parkway NW, Suite 300
Boca Raton, FL 33487-2742

First issued in paperback 2019

ISBN-13: 978-0-8493-7255-1 (hbk)
ISBN-13: 978-0-367-38976-5 (pbk)

Visit the Taylor & Francis Web site at
http://www.taylorandfrancis.com

and the CRC Press Web site at
http://www.crcpress.com

Library of Congress Cataloging-in-Publication Data

Loke, Seng.
    Context-aware pervasive systems : architectures for a new breed of applications / Seng Loke.
       p. cm.
    Includes bibliographical references and index.
    ISBN 0-8493-7255-0 (alk. paper)
    1. Ubiquitous computing. 2. Mobile computing. 3. Software architecture. I. Title.

QA76.5915.L65 2006
005.1'2--dc22
                                  2006049923

DEDICATION

To my Creator who knows all about aware systems, living and nonliving.

In memory of my dad (1936–1998) who is now much more aware
of important things and the Lord Jesus Christ.

PREFACE

Context-aware mobile computing has been a topic for research since one of the earliest Ph.D. dissertations appeared on the topic in 1994. Recently, context-aware computing has enjoyed remarkable attention from researchers in diverse areas such as distributed computing and human–computer interaction. Such aware systems have become one of the most exciting concepts in early 21st-century computing, fueled by recent developments in pervasive computing (i.e., mobile and ubiquitous computing) including new computers worn by users, embedded devices, smart appliances, and sensors surrounding users and varieties of wireless networking technology. Software and hardware systems that are ubiquitous and aware of users, and their physical and virtual context (e.g., environment and circumstances), and can respond intelligently to what is perceived is an exciting, if not increasingly vital, addition to daily life and work. Whereas the idea of context has been studied in logic and the meaning of natural language sentences, the notion of context is being revisited in mobile and ubiquitous computing work. The experience economy has taught us that experience matters, and context awareness is a key idea for providing new experiences with devices, appliances and software systems, and automatic behaviors for convenience and innovative applications.

This book is a gentle introduction to a new breed of computer applications termed *context-aware pervasive systems*, and attempts to provide architectural blueprints for building context-aware behavior into applications. The book reviews the anatomy of context-aware pervasive applications, including:

- Context-aware mobile services
- Context-aware devices, appliances, and smart things
- The integration of context-aware computing with software agents and the Web

- The use of context awareness for addressing, and communication between, people, devices, and software agents
- Context-aware controlled sensor networks
- Context-aware security frameworks
- Context awareness via mirror worlds

In this book we aim to capture general design principles and architectures for context-aware applications. These applications are certainly not exhaustive and only serve to illustrate the usefulness and potential of context awareness in mobile and ubiquitous systems, and the range and diversity of context-aware behaviors, to the extent that can be done within one book. The book also highlights the notion of *mirror worlds* (a term I believe originated in Gelernter's book) and its interesting applicability to building aware systems, and discusses declarative approaches to constructing such systems. I have often used examples from my own work to illustrate the concepts presented here, perhaps not surprisingly; where appropriate, I have noted work by others in the area. Although related work has been surveyed, there is work which I have left out, given the extent of activity in the area. I trust that readers will find the extensive set of references useful, and use this book as a platform to further explore the area.

Context-aware pervasive computing is still an area of active research, and we will indeed develop a deeper understanding of such systems, better techniques, and architectures of greater generality. Hence, one faces a dilemma about writing a book too early for an area that is still growing and in many ways changing. Nevertheless, I feel that there is a need for such a book, serving as a timely and relevant introduction to the emerging breed of context-aware systems, and presenting an initial step toward bringing together in one volume architectures and principles — as they relate to the applications covered — of such systems, providing material already in use by practitioners and enthusiasts in an exciting field.

One of the titles initially selected for this book was *Inside Aware Systems: Introducing the Software Architectures of a New Breed of Applications*. Dropping the "context" from "context-aware" is a move toward a more general concept, which perhaps this book can help develop.

Readers can become acquainted here with an increasingly important new breed of software and their implications and possibilities, even if they are not experts in the field or directly work in the area. Students and researchers new to the area can quickly obtain familiarity with key ideas and concepts of the topic, all in one place, acquiring a framework by which to understand related work and perhaps to start working in the area. Practitioners can take the designs and architectures presented and implement their own versions of the systems, adding their own features or improvising as their applications require. Researchers from other areas can find application of their own expertise within the area of aware systems, based on the abstract architectures presented here. Nontechnical readers, skipping over the technical material, will still be able to gain an appreciation for the ideas and concepts within the area.

ACKNOWLEDGMENTS

This book existed in an early form in bits and pieces, distributed in several papers, and published in a number of conference and workshop proceedings and journals over the past few years. The substance of this book is not the work of one person. I would like to thank the many graduate students and colleagues who worked with me on these topics and continue to do so. To mention a few, Evi Syukur, Thin Thin Naing, and Angel On Kei Tam worked on mobile services; Sutardi on the context-aware mobile phone prototype; Shonali Krishnaswamy, Andiwijaya Sumartono, and Su Hui Chuah on the CALMA agent-based Web service framework; and Amir Padovitz and Arkady Zaslavsky on context-aware messaging and context modeling. Suan Khai Chong and Shonali worked with me on context-aware sensors (Chapter 7 is, in fact, based on a draft paper that Khai wrote); Yong Jin Sim on a mutual awareness model for devices (mentioned in Chapter 9), and Pravin Shetty explored context-aware security with me. I also would like to thank Toan Phung, Budi Halim, and Jonathan Yu for helping with the location-based agents' prototype and contributing ideas over their summer vacation. The students helped turn what is in our imagination into real working systems on real devices, and I thank them for giving me the satisfaction of seeing my imagination crystallized. The Mobility, Agents and Pervasive Systems Group at Monash University, Caulfield, Melbourne, was a wonderful environment in which to think, pursue, and realize new ideas and concepts.

I completed this book at La Trobe University's Department of Computer Science and Computer Engineering, and I would like to acknowledge the conducive and supportive department that it is. It is both encouraging and invigorating to be in an environment excited about pervasive computing.

ABOUT THE AUTHOR

Dr. Seng Loke is currently a senior lecturer in the Department of Computer Science and Computer Engineering at La Trobe University, Melbourne, Australia, and is also an honorary associate of the Center for Distributed Systems and Software Engineering at Monash University. He was previously a senior fellow at Monash University. He has published more than 130 papers as book chapters and in journals, conferences, and workshops. He co-leads the Pervasive Computing Group at La Trobe.

CONTENTS

1

WHAT IS CONTEXT-AWARE BEHAVIOR?

"Good morning, Smith! Coffee is boiling and your E-newspaper has been downloaded," a pleasant voice greets you as you walk into the kitchen from the bedroom. As you sit in your car, the seats, mirrors, and steering wheel are automatically adjusted — your son had been using your car last night. In the office, as you sit on the chair in front of your desk, the height of the chair is automatically adjusted — your short colleague had been sitting in your chair, again — and you are automatically logged in to your computer. At lunch time, while in the mall, you are greeted with messages appearing on your smartphone from restaurants serving your favorite Chinese noodles, located no more than 200 m away. Only one's imagination could limit the possibilities of systems that can be aware of people's situations or their contexts and do things for them. *The Economist* discusses the sentient office[1] containing technologies that are convivial not only in the office but also at home, "using information about where you are in your home to control the lighting or ensure that only your nearest telephone rings or that the television program you are watching follows you from screen to screen as you move between rooms." Movies such as *Artificial Intelligence*, *I, Robot*, and *The Matrix* contain futuristic scenarios that, if not going even further beyond, touch the boundaries of what is achievable. But it is not only robots which can have the ability to sense, reason, and act — almost anything can, from the coffee cup, television, soft toys, and the house to the bed.

---

[1] http://www.economist.com/science/tq/PrinterFriendly.cfm?Story_ID=1841108.

1.1 CURRENT COMPUTING TRENDS: FROM THE VIRTUAL TO THE PHYSICAL

Programmers have tremendous power to exert their imagination to create not only office applications but also virtual worlds. From games programming to avatars, virtual environments have become widespread. Recently, increasing attention is being placed on augmenting the physical environment using information technologies. Pattie Maes' *intelligence augmentation*[2] is a phrase used to describe how the physical world can be enriched with intelligent behavior. An area can become an avenue where technologies "pile up" and become interactive (McCullough, 2005) and somewhat aware of their inhabitants, leading to the notion of smart environments, the creation of which has been documented by Cook and Das (2004).

Computer systems that surround, pervade, and intelligently serve people in pleasant and unobtrusive ways provide a vision that has seized the imagination of many. These computer systems might not fit the traditional image of a computer sitting on a desk. Articulation of this vision and its variations has resulted in buzz phrases such as *ubiquitous computing, pervasive computing, invisible computing, the disappearing computer, proactive computing, autonomic computing, ambient intelligence*, and *sentient computing*. But these are not merely phrases; much technology lies beneath the new vocabulary and are being developed under these topics.

- *Ubiquitous computing* (Weiser, 1991)[3] refers to the collective use of computers available in the physical environment of users, perhaps embedded in a form invisible to users. This is the vision of the late Mark Weiser from Xerox PARC for putting computers out into everyday living environments, instead of representing the everyday living environment in the computer.
- *Pervasive computing* refers to the vision of devices or computers pervading lives, as IBM Chairman Lou Gerstner once described: "… A billion people interacting with a million E-businesses with a trillion intelligent devices interconnected." Pervasive computing can be viewed as a combination of mobile computing (use of computers worn on or carried by users) and computers embedded in the fixed environment and so can be understood as another term for ubiquitous computing.
- *Invisible computing* (Norman, 1998; Borriello, 2000) refers to the use of computers in such a way that the task is focused on and facilitated, without too much focus on the tool (i.e., the computer

---

[2] http://www.kurzweilai.net/meme/frame.html?main=/articles/art0264.html.

[3] See also http://www.ubiq.com/hypertext/weiser/UbiCACM.html and http://en.wikipedia.org/wiki/Ubiquitous_computing.

system) itself. Often, we simply want a task to get done (e.g., get a message to someone), and we might not want to focus on, or wrestle with, the software or the hardware that enables the task. If there is a cognitive burden on the user in employing a tool to perform a task, the more this burden is laid on the task rather than the tool, the better. The overlap with the ubiquitous computing vision is clear; the many computers in the everyday environment that the user might interact with are merely the tools to perform tasks and so should be given only as much attention as a tool deserves.

■ The European Union-funded *disappearing-computer* initiative[4] aims to create artifacts commonly seen or used in everyday life with computational capabilities (using some combination of hardware and software) and the ability to work together to produce new behaviors. It also looks at people's experiences with environments comprised of such artifacts. The initiative has a number of projects, including those that explore wearable computer devices and that study how a colocated collection of gadgets might cooperate.[5]

■ *Proactive computing* (Tennenhouse, 2000) refers to a focus away from interactivity to computers anticipating user needs and taking action on users' behalf. The idea is that human attention devoted to interaction can be reduced so that users can focus on higher-level tasks.

■ Another recent topic is *autonomic computing* (Horn, 2001), which is about building systems that can self-monitor, self-heal, and self-configure. Proactive and autonomic computing overlap, in that they both relate to ubiquitous computing systems and could utilize context information about the environment of the system and users to operate or make decisions. However, they differ in their emphasis on the kind of system behavior to be achieved. A deeper comparison between autonomic and proactive computing is presented by Want et al. (2003).

■ *Ambient intelligence* (Marzano and Aarts, 2003) builds on ubiquitous computing and intelligent user interfaces to obtain greater user friendliness and efficient services for users. To quote from *Ercimnews*[6] "In this vision, people will be surrounded by intelligent and intuitive interfaces embedded in everyday objects around us and an environment recognizing and responding to the presence

---

[4] See the main Web site for the initiative: http://www.disappearing-computer.net/.

[5] http://www.extrovert-gadgets.net/.

[6] http://www.ercim.org/publication/Ercim_News/enw47/intro.html.

of individuals in an invisible way by year 2010." Certainly, context-aware computing overlaps with the ambient intelligence vision.

■ *Sentient computing* (Hopper, 2000)[7] refers to systems "using sensors and resource status data to maintain a model of the world which is shared between users and applications." Because such systems try to build a model of a part of the world from sensory information about the user's circumstances and environment, the idea is very much suggestive of, if not synonymous with, context-aware computing but with an emphasis on the world model.

From the brief description of the topics mentioned, the reader can easily observe the overlap in the technologies they aim to create. Ubiquitous, pervasive, disappearing, invisible, proactive, autonomic, and sentient computers exhibiting intelligent behavior and surrounding the user are a current goal in computing. An aspect of this vision is context-aware behavior in a pervasive computing setting. The next section explores further the notion of context and context awareness.

1.2 CONTEXT, CONTEXT AWARENESS, AND SITUATIONS

The notion of context has been observed in numerous areas, including linguistics, philosophy, knowledge representation and problem solving in the field of artificial intelligence, and the theory of communication (Akman, 2002; Bouquet et al., 2003; McCarthy, 1993; Brezillon, 2003). In most work of this nature, context is a key notion (e.g., treated as first-class objects in a logic), and logics have been developed to enable assertions to be made about contexts and allow context to be reasoned about explicitly in knowledge-based systems.

What is context? The *Free Online Dictionary of Computing*[8] defines *context* as "that which surrounds, and gives meaning to, something else." This definition can be specialized to the application at hand. Whether that "something" is an assertion in a logic, a person, an utterance, or a computer system, with a suitable definition for "meaning," the intuition captured by the word *context* serves its purpose. Schilit et al. (1994) define context from the perspective of distributed and mobile computing, where a person is that "something," and *context* refers to information about a person's proximate environment, such as location and identities of nearby people and objects.

Dey (2001) gives an operational definition of context, which turns out to be very useful in practice and suitable for pervasive computing: "Context

---

[7] See http://en.wikipedia.org/wiki/Sentient_computing.

[8] http://www.dictionary.com.

is any information that can be used to characterize the situation of an entity. An entity is a person, place, or object that is considered relevant to the interaction between a user and an application, including the user and applications themselves." There has been much work in identifying what such information can be, the structure of the information, how to represent such information, and how to exploit context in specific applications. Contexts can include information such as location (e.g., of people or objects), time, execution state of applications, computational resources, network bandwidth, activity, user intentions, user emotions (Picard, 1997), and conditions of the environment (Dey, 2001). Models might be used for more complex kinds of contextual information, such as location models,[9] world models (e.g., Lehmann et al., 2004), and activity models[10] (e.g., Muhlenbrock et al., 2004; Koile et al., 2003; Tapia et al., 2004).

Indeed, there is tremendous diversity in what can be context, and the way context can be acquired and modeled is an avenue of much research. Recent workshops[11] have focused on just this topic. To address the challenges in representing, structuring, managing, and using context, various knowledge-representation formalisms and techniques have been applied, including ontologies (Chen et al., 2004; McGrath et al., 2003; Wang et al., 2004; Matheus et al., 2003) that provide concepts for describing context and enable reasoning with and reuse of contextual information, first-order logic theories (Katsiri and Mycroft, 2003; Ranganathan and Campbell, 2003), and conceptual graphs (Peters and Shrobe, 2003). One needs to consider what aspects of the physical world to sense for a given application to operate. The ontologies provide a vocabulary of concepts with which to express contexts, and formal mathematical representations enable automatic reasoning with the knowledge on the computer.

Pervasive computing utilizes contextual information about the physical world. This implies a number of important concerns related to the connection of sensor information to context-aware pervasive computing,

---

[9] See http://research.microsoft.com/workshops/UbiLoc03/.

[10] Typically, *activity* refers to some action or operation undertaken by a human being, such as bathing, studying, preparing breakfast, and playing table tennis, and so differs from one situation to another. Perhaps one could conceive of a person in the state of preparing breakfast as "a situation." However, in general, *activity* and *situation* are clearly not interchangeable, and we consider activity as a type of contextual information that can be used to characterize the situation of a person (e.g., that "breakfast is being prepared" means the person is busy or has just woken up).

[11] Examples of workshops on context are the Workshop on Context Modeling and Reasoning (CoMoRea 2004), Workshop on Modelling and Retrieval of Context (MRC 2004) (http://mrc2004.wysart.de/), and Workshop on Advanced Context Modelling, Reasoning, and Management 2004.

including (1) what can be feasibly sensed, (2) the best way to acquire sensor information, and (3) how to reason with sensor information to infer context. In fact, any information which can be practically obtained via sensors can be used as context, including the emotional states of users and movements.

When the entity is an artifact instead of a person, we have context-aware artifacts. From buildings to dolls, one can add such sensors to endow these artifacts with the ability to act on sensed information about the physical world. Making such artifacts "aware" enables automatic behaviors without users' direct intervention and can enhance the function of the artifact or add aesthetic value, both of which can differentiate an artifact from others in the marketplace.

Closely related to the notion of context is the notion of situation. The relationship between context and situation is illustrated in Dey's operational definition cited earlier. A definition of *situation* from the *American Heritage Dictionary*[12] is as follows: "The combination of circumstances at a given moment; a state of affairs." Besides describing *context*, Dey (2001) also defines *situation* as "a description of the states of relevant entities." So, the idea is to aggregate (perhaps varieties of) context information to determine the situation of the entities (relevant to an application). In this sense, we can view situation as being at a higher level of abstraction than context.

Similar to context, the notion of situation has been explored in artificial intelligence (AI), philosophy, and linguistics, such as in situation theory (Barwise and Perry, 1983) and situation calculus. This perspective considers the primacy of the situation abstraction and has noted that an agent (e.g., a human) is able to individuate a situation. According to Devlin (1991), a situation is a "structured part of reality that it (the agent) somehow manages to pick out" by "direct perception of a situation, perhaps the immediate environment, or thinking about a particular situation," and "individuation of a situation by an agent does not (necessarily) entail the agent being able to provide an exact description of everything that is and is not going on in that situation."

The notion of situation can be useful in pervasive computing in that the situation abstraction allows the modeler or application designer to effectively "carve the world up" into manageable pieces, which can then be recognized by the system via its sensors. It might also be possible to compose such pieces to construct more complex models of situations, as we shall see later in the book. A system can become aware of the context of a given set of entities and then guess what situations those entities are

---

[12] Accessed from http://www.dictionary.com.

in, or detect a change of situation. Machine understanding of situations is a goal of gathering contextual information.

1.3 WHEN SYSTEMS BECOME CONTEXT AWARE

Quoting from Schilit (1995): "... *Context-aware software* adapts according to the location of use, the collection of nearby people and objects, the accessible devices, as well as changes to those objects over time. A system with these capabilities surveys the computing environment and reacts to changes to that environment." Context-aware pervasive computing is a study of pervasive computer systems (a combination of hardware and software)[13] that are aware of context and can automatically adapt and respond to such context. Context awareness enables the system to take action automatically, reducing the burden of excessive user involvement and providing proactive intelligent assistance.

Saha and Mukherjee (2003) consider such use of perceptual information about the environment an essential ingredient of pervasive computing systems,[14] distinguishing them from traditional computing. Such context-aware pervasive systems have wide-ranging applications, including context-aware information retrieval (Brown and Jones, 2001), reminder systems (Rhodes, 1997), context-aware mobile services and electronic tour guides (Abowd et al., 1997), sentient objects (Fitzpatrick et al., 2002; Biegel and Cahill, 2003), sentient cars (Vidales and Stajano, 2002), sentient buildings,[15] context-aware response to emergencies,[16] and intelligent context-aware environments (Shafer et al., 2001), with potential benefits to society, ranging from proactive automated healthcare (Bardram, 2004)[17] and effective E-commerce (Jin and Miyazawa, 2002) to more effective military systems (van der Poel, 2002) and safer cars. Indeed, one can also utilize context awareness in security warning systems in the home or public spaces. Many future computing systems will advantageously be context aware, and context-aware systems are becoming increasingly important, receiving worldwide attention from academia and industry.

From robots to automatic surveillance systems, one could think of many existing systems that might be viewed as having "awareness." However, there is a rethinking of what it means for systems to be context aware, and research continues on general principles and architectures, as

---

[13] We keep in mind that the systems might not look like a typical computer or be recognized as such by users.

[14] Pervasive computing combines mobile computing and ubiquitous computing (Satyaranayan, 2001).

[15] See http://www.wikipedia.org/wiki/Sentient_computing.

[16] See http://guir.berkeley.edu/projects/emergency/.

[17] See http://www.pervasivehealthcare.dk/projects/.

well as specialized designs. There is also the application of this idea to many different items, yielding "context-aware <object>" where <object> could, in theory, be almost anything. It is not only that new kinds of artifacts and computer applications are made possible by context awareness, but existing desktop applications can usefully be made context aware. Traditional applications, which normally run on a desktop environment, can be driven or controlled by contextual information about users. For example, Windows Media Player can be made context aware — starting and stopping according to who is in the room or the mood of its occupants. Also, as mentioned, an artifact can be made context aware in different ways, i.e., being aware of different kinds of contextual information and reacting to situations in different ways. Such awareness is perhaps one aspect of objects when they start to think (Gershenfeld, 1999).[18]

A new breed of computer systems is emerging that is context aware and so different from traditional computer applications of past decades. These systems will be more sensitive to the real world and attuned to the purposes for which they have been developed.

1.4 AN OVERVIEW OF THIS BOOK

This book looks at the architectures and internals of examples of context-aware pervasive systems, illustrating how they work and how they can be designed. Chapter 2 presents a generic abstract architecture for context-aware pervasive systems, noting its core elements. Such an architecture represents a general blueprint for the examples that follow (perhaps implicitly) in the book — each example adding its own specific features. The chapter also discusses the different areas of computing that building such systems tend to integrate, including software engineering, AI reasoning, sensor networks, and Internet computing.

Mobile services, appliances and smart devices, software agents, electronic communication, sensor networks, and security frameworks are topics worthy of their own books, and books have indeed been written on these topics. In Chapter 3 through Chapter 8, we provide a brief introduction to these topics before showing how context-aware behaviors can be explored for these topics.

Chapter 3 focuses on context-aware mobile services. Internet-based digital services have become increasingly important to daily life, and early ideas of context-aware computing have been explored in connection with mobile devices. Indeed, mobility of persons and devices gives rise to a change of location. Not only is location a key type of contextual information, but a change in location often comes with a change in the

---

[18] See Web site on the "Things That Think Consortium," http://ttt.media.mit.edu/.

environment, yielding a corresponding change in other types of contextual information. Where a person or device is situated can speak volumes about a person's situation as we shall see. Location and place do matter.

Chapter 4 discusses context-aware devices, appliances, and smart things. Pervasive computing encourages the imagination — what if computational ability is added to the ordinary mundane objects we see everyday? Perhaps outrageous is a kettle that senses you (or whoever is in the kitchen) and is able to make small talk with you (or the person it identifies), giving you some good news or relaxing words while it boils the coffee during your break. Is such ability feasible? What are some efforts in this area? What software architectures are involved? What is the scope of the computational abilities required? These are some questions the chapter seeks to answer.

Intelligent software agents (Wooldridge, 2002) are a recent area of research that has spread like wildfire through computing labs throughout the world. Although a hot topic in itself, there is an interesting overlap between the essence of context-aware systems and the concept of the intelligent software agent. Which metaphor to use in designing effective computer systems is a question to be answered. Chapter 5 introduces intelligent software agents and discusses an example of context-aware agents used with the Web.

In the real world, context is key to identifying or addressing someone or something and in sending and receiving messages (whatever form they may be). We are very much embedded within a world which gives meaning to our communication and helps us identify people and objects and direct our messages. For example, we can talk about a person next to someone else we know as a means of identifying the person or about putting off receiving a communication (e.g., a phone call) because of current circumstances. Chapter 6 discusses the use of context awareness for addressing, and communication between, people, devices, and software agents.

Sensor networks are in many ways to pervasive systems what the five senses are to a human being. Sensor networks are, nevertheless, a complex topic in itself, from networking to query processing. We focus on a specific aspect of sensor networks, i.e., power management. Chapter 7 considers how sensors themselves can be aware of their own situations and explores an application of this idea to power saving, one of the key issues in the use of sensor networks.

Distributed computing has added a new range of security issues to traditional desktop computing. Mobile computing has gone even further, bringing in further new security concerns. Chapter 8 looks at the use of physical context for controlling and enhancing security in pervasive computing environments.

Chapter 9 and Chapter 10 consider two useful perspectives on building context-aware systems and environments. Chapter 9 describes the notion of mirror worlds and explains how implementing such mirror worlds is an approach to building aware systems, echoing the ideas in sentient computing mentioned earlier in the context of a world model that is utilized and maintained. Chapter 10 discusses design perspectives based on a declarative programming language paradigm.

REFERENCES

Abowd, G., Atkeson, C., Hong, J., Long, S., Kooper, R., and Pinkerton, M., Cyberguide: a mobile context-aware tour guide, *ACM Wireless Networks* 3, 421–433, 1997.

Akman, V., *Context in Artificial Intelligence: A Fleeting Overview* (English version of Contesti in intelligenza artificiale: una fugace rassegna), in Penco, C., Ed., *La Svolta Contestuale*, McGraw-Hill, Milano, 2002.

Bardram, J., Applications of context-aware computing in hospital work — examples and design principles. *Proceedings of SAC '04*, March 14–17, Nicosia, Cyprus, 2004, pp. 1574–1579.

Barwise, J. and Perry, J., *Situations and Attitudes*, Cambridge, MA: MIT-Bradford, 1983.

Biegel, G. and Cahill, V., Sentient Objects: Towards Middleware for Mobile Context-Aware Applications. European Research Consortium for Informatics and Mathematics, ERCIM News No. 54, July 2003.

Borriello, G., The challenges to invisible computing, *IEEE Computer*, 33(11), 123–125, 2000.

Bouquet, P., Ghidini, C., Giunchiglia, F., and Blanzieri, E., Theories and uses of context in knowledge representation and reasoning, *Journal of Pragmatics*, 35(3), 455–484, 2003, Elsevier Science.

Brezillon, P., Representation of procedures and practices in contextual graphs, *The Knowledge Engineering Review*, 18(2), 147–174, 2003, Cambridge University Press.

Brown, P.J. and Jones, G.J.F., Context-aware retrieval: exploring a new environment for information retrieval and information filtering, *Personal and Ubiquitous Computing Journal* 5(4), 253–263, 2001.

Chen, H., Finin, T., and Joshi, A., An ontology for context-aware pervasive computing environments, *The Knowledge Engineering Review (special issue on Ontologies for Distributed Systems)* 18(3), 197–207, 2004.

Cook, D., and Das, S., *Smart Environments: Technology, Protocols and Applications*, Wiley-Interscience, U.S.A., 2004.

Devlin, K.J., Situations as mathematical abstractions, in J. Barwise, J.M. Gawron, G. Plotkin, and S. Tutiya (Eds.), *Situation Theory and its Applications*, CSLI, Stanford, CA, 1991, pp. 25–39.

Dey, A.K., Understanding and using context, *Personal and Ubiquitous Computing Journal*, 5(1), 5–7, 2001.

Fitzpatrick, A., Biegel, G., Clarke, S., and Cahill, V., Towards a Sentient Object Model. Position Paper Workshop on Engineering Context-Aware Object Oriented Systems and Environments (ECOOSE), Seattle WA, November 2002, available at http://www.dsg.cs.tcd.ie/~biegelg/research/publications/biegel-som.pdf.

Gershenfeld, N. *When Things Start to Think*, Henry Holt and Company, New York, 1999.

Hopper, A., The Royal Society Clifford Paterson Lecture, 1999 — Sentient Computing, Philosophical Transactions, Royal Society London, Vol. 358, August 2000, pp. 2349–2358.

Horn, P., Autonomic Computing: IBM's Perspective on the State of Information Technology, October, 2001, available at http://www.research.ibm.com/autonomic/manifesto/autonomic_computing.pdf.

Jin, L. and Miyazawa, T., MRM server: a context-aware and location-based mobile e-commerce server, *Proceedings of the 2nd International Workshop on Mobile Commerce*, USA, 2002, ACM Press, pp. 33–39.

Katsiri, E. and Mycroft, A., Knowledge representation and scalable abstract reasoning for sentient computing using first-order logic, in *Proceedings of the Challenges and Novel Applications for Automated Reasoning (CADE-19 Workshop)*, July 2003, available at http://www.cl.cam.ac.uk/users/am/papers/nads03.pdf.

Koile, K., Tollmar, K., Demirdjian, D., Shrobe, H., and Darrell, T., Activity zones for context-aware computing, in *Proceedings of the 5th International Conference on Ubiquitous Computing (UBICOMP 2003)*, 2003.

Lehmann, O., Bauer, M., Becker, C., and Nicklas, D., From home to world — supporting context-aware applications through world models, in *Proceedings of the 2nd IEEE Annual Conference on Pervasive Computing and Communications (PERCOM'04)*, IEEE Computer Society, 2004.

Marzano, S. and Aarts, E., *The New Everyday View on Ambient Intelligence*, Uitgeverij 010 Publishers, 2003.

Matheus, C., Kokar, M., and Baclawski, K., A core ontology for situation awareness, in *Proceedings of FUSION*, July 2003, pp. 545–552, available at http://vistology.com/papers/FUSION03.pdf.

McCarthy, J., Notes on formalizing contexts, in Ruzena Bajcsy, Ed., *Proceedings of the 13th International Joint Conference on Artificial Intelligence*, Morgan Kaufmann, San Mateo, CA, 1993, pp. 555–560.

McCullough, M., *Digital Ground*, MIT Press, 2005.

McGrath, R.E., Ranganathan, A., Campbell, R.H., and Mickunas, M.D., Use of Ontologies in Pervasive Computing Environments, Technical Report, UIUCDCS-R-2003-2332 ULU-ENG-2003-1719, April 2003, available at http://mummy.intranet.gr/includes/docs/MUMMY-D11y1-ZGDV-CtxtAwr-v02.pdf.

Muhlenbrock, M., Brdiczka, O., Meunier, J.-L., and Snowdon, D., Learning to detect user activity and availability from a variety of sensor data, in *Proceedings of the 2nd IEEE Annual Conference on Pervasive Computing and Communications (PERCOM'04)*, IEEE Computer Society, 2004.

Norman, D., *The Invisible Computer*, MIT Press, 1998.

Peters, S. and Shrobe, H.E., Using semantic networks for knowledge representation in an intelligent environment, in *Proceedings of the 1st IEEE Annual Conference on Pervasive Computing and Communications (PERCOM'03)*, 2003, IEEE Computer Society, Washington, D.C., pp. 323–329.

Picard, R.W., *Affective Computing*, MIT Press, 1997.

Ranganathan, A. and Campbell, R.H., An infrastructure for context-awareness based on first order logic, *Personal and Ubiquitous Computing Journal*, 7, 353–364, 2003.

Rhodes, B., The wearable remembrance agent: a system for augmented memory, *Proceedings of the 1st International Symposium on Wearable Computers*, Cambridge, MA, 1997, pp. 123–128.

Saha, D. and Mukherjee, A., Pervasive computing: a paradigm for the 21st century, *IEEE Computer,* 25–31, 2003.

Schilit, B.N., Adams, N.I., and Want, R., Context-aware computing applications, in *Proceedings of the Workshop on Mobile Computing Systems and Applications,* December, 1994, IEEE Computer Society, pp. 85–90.

Schilit, B.N., A System Architecture for Context-Aware Mobile Computing, Ph.D. thesis, Columbia University, 1995, available at http://seattleweb.intel-research.net/people/schilit/schilit-thesis.pdf.

Shafer, S., Brumitt, B., and Cadi, J.J., Interaction Issues in Context-Aware Intelligent Environments. *Human Computer Interaction* 16(2–4), 2001, available at http://www1.ics.uci.edu/~jpd/NonTradUI/SpecialIssue/shafer.pdf.

Tapia, E.M., Intille, S.S., and Larson, K., Activity recognition in the home using simple and ubiquitous sensors, *Pervasive 2004, LNCS 3001,* Springer-Verlag, Germany, 2004, pp. 158–175.

Tennenhouse, D.L., Proactive computing, *Communications of the ACM,* 43(5), 43–50, 2000.

van der Poel, B., Context-Aware Rule-Based Data Distribution Algorithms and Methods for Pervasive Computing. M.Sc. thesis, Delft University of Technology, August 2002.

Vidales, P. and Frank Stajano, F., The sentient car: context-aware automotive telematics, *Proceedings of First IEEE European Workshop on Location Based Services (LBS-2002),* London, also appeared as a poster at Ubicomp 2002, available at http://www-lce.eng.cam.ac.uk/~fms27/papers/2002-VidalesSta-car-lbs.pdf.

Wang, X.H., Gu, T., Zhang, D.Q., and Pung, H.K., Ontology based context modeling and reasoning using OWL, in *Proceedings of the Workshop on Context Modelling and Reasoning (CoMoRea'04) at the 2nd IEEE International Conference on Pervasive Computing and Communications,* IEEE Computer Society Press, 2004.

Want, R., Pering, T., and Tennenhouse, D., Comparing autonomic and proactive computing, *IBM Systems Journal,* 42(1), 2003.

Weiser, M., The computer for the twenty-first century, *Scientific American,* September, 1991, pp. 94–10.

Wooldridge, M., *An Introduction to MultiAgent Systems,* John Wiley & Sons, New York, 2002.

2

THE STRUCTURE AND ELEMENTS OF CONTEXT-AWARE PERVASIVE SYSTEMS

Examples of the behavior of context-aware pervasive systems in the previous chapter suggest that there are commonalities among these systems and that they need to be specifically designed and constructed to achieve these behaviors. The range, diversity, and sophistication of context-aware applications have continued to increase (Chen and Kotz, 2000; Mitchell, 2002), yet one can notice reusable concepts and software architectures relevant to new applications to be developed.

This chapter first introduces analogies by which we can understand aware systems and describes the basic elements of a context-aware pervasive system, from sensors and modeling to reasoning techniques. Then, a generic abstract architecture for context-aware pervasive systems is presented, and a brief review of selected infrastructures, middleware, and toolkits for context-aware pervasive computing is given.

2.1 ANALOGIES

We are able to perceive the world through the five senses of touch, taste, smell, sight, and hearing. Our brain is able to make use of whatever impinges on our sense organs so that we experience sensation. Our reaction to stimuli can be almost immediate, such as the knee-jerk effect, or delayed, perhaps taking place after long and careful planning. Our experience of the world can be viewed as either direct via the senses, or indirect via reasoning, with the combination of knowledge that we already have and new information gathered via the senses.

Insects have senses and are capable of reacting to stimuli. Although they do not have the complex brain of a human being, they are able to show interesting behaviors. For example, ants are able to show collective behavior when finding food. Through leaving pheromone trails for one another, ants organize themselves into groups, which gather information from food sources efficiently; collective intelligent behavior emerges.

Intelligent software agents are an emerging class of software systems that are proactive, autonomous, communicative (with people and other agents), and adaptive. These software systems are also situated in their environment and react and respond to stimuli from their environment. Whereas physical-world properties can be converted into data in the computer and used by these agents, such agents typically inhabit a software environment. Multiagent systems which involve a combination of such agents working together provide a software paradigm for building large, complex distributed systems. One popular use of the term *agent* is as a metaphor for software that does something on behalf of another, and the terms *information agents* and *internet softbots* (Etzioni, 1997) have been used to describe software that search, monitor, filter, and process information in the environment of the World Wide Web. Related to software agents are robots that can sense the physical environment and respond as appropriate, performing human-like behaviors, perhaps more efficiently than humans.

Humans, insects, intelligent software agents, and robots are clearly different categories of entities but, interestingly, one can see that, at a high level of abstraction and from the right perspective, they share something in common with context-aware pervasive systems — the ability to sense and respond to stimuli, sometimes with behavior that we would describe as *intelligent*. What is meant by "sensing" and the kind of response to stimuli might differ across categories and between specific entities within these categories. Nevertheless, we gain an insight into the nature of these entities through this abstract view. Context-aware pervasive systems have their unique emphasis on identifying, understanding, and exploiting the context of entities and on architectures that involve pervasive computing technologies of myriad interconnected devices and components surrounding these entities. Context-aware pervasive systems might also be built from augmenting everyday objects and traditional software systems with sensing and stimuli-responsiveness.

2.2 THE ELEMENTS OF A CONTEXT-AWARE PERVASIVE SYSTEM

A context-aware pervasive system can be viewed as having three basic functionalities: sensing, thinking (metaphorically), and acting. Systems can vary in sophistication in each of these functionalities. Some systems might include complex sensors but perform little reasoning before acting. Others might utilize little sensing but perform much deliberation before acting. Also, these functionalities can be realized in a centralized or a distributed architecture over one or more physical devices.

2.2.1 Sensing

Sensors, biological or nonbiological, provide a means to acquire data or information about the physical world or some aspect of the physical world. Such knowledge can be used by a computer system to determine actions most appropriate to the physical situation at hand. A combination of multiple sensors can give even more information for the computer system to reason with, providing a more comprehensive view of the physical world. A computer program can only normally compute with the inputs it is given and, traditionally, such inputs are provided manually by users. Such inputs can be provided by sensors, which can be then viewed as providing a bridge between the physical world and the virtual world of the computer program.

What kind of information can be sensed? A large variety of sensors have been developed, including light sensors, temperature sensors, smoke detectors, motion sensors, and touch sensors. However, there are many more devices that can be viewed as sensors, including microphones, pressure gauges, and the computer clock. In Schmidt's Ph.D. dissertation (Schmidt, 2002), sensing technologies listed include those for light and vision, audio, movement and acceleration, location and position, magnetic field and orientation, proximity, touch and user interaction, temperature, humidity and air pressure, weight, motion, gas and smell (e.g., electronic noses), and biological signals (e.g., measuring heart rate, skin resistance, muscle tension, blood pressure, etc.). Klein (1999) mentions radiometers, radars, and infrared sensors.

An upcoming technology that has been receiving widespread attention is radio frequency identification (RFID) tags or smart labels (Lahiri, 2005). RFID tags can be read from or, in some cases, written to, using an RFID reader with energy from a radio frequency field. RFID tags can store 64 to several thousand bits of data. One can tag everyday objects such as documents, books, or pill bottles with RFID tags and have a reader detect the presence or absence of one or more of these objects (Want et al.,

1999). When the RFID tag or the RFID-tagged object is within several centimeters of the reader (or in some RFID technologies, further), the object is detected, and when the object is moved further away, the absence of the object is detected. In a supermarket, one could explore RFID-tagged grocery items for automatically identifying and adding up the price of items at checkout. A basketful of tagged items can also be detected and analyzed.

RFID tag systems can be viewed as a sensor that provides information about the relative physical position of these objects (i.e., between RFID readers and tags). An interesting application of RFID tags is to construct a model of the configuration of a number of physical parts; for example, given a collection of tagged parts such as those in a car, one can identify missing parts or parts in the wrong places.

Location is a widely used type of context in applications. Location-based applications abound, from finding the nearest ATM, navigating through unknown streets, and contextualizing Web pages to determining which room a particular person is in within a building (Jagoe, 2002). Depending on the positioning technology used, location information can be represented in different ways — including coordinates in some coordinate system, human-understandable labels (perhaps translated from the coordinates) and in relative terms (e.g., is the object near some other object?) — and in different granularities.

The granularity of location information might be within centimeters, meters, or greater. For example, global positioning system (GPS) use satellites (perhaps assisted by ground stations) to determine the position of a GPS receiver within an accuracy of several meters (Andersson, 2001). Mobile phone networks can determine the location of an individual within a suburb or town (around 150 to 300-m accuracy) in the case of an emergency call.

On a smaller scale, such as within an area of a radius of the several hundreds of kilometers, serviced by wireless local area networks (WLANs), positioning technologies such as Ekahau[1] use triangulation with multiple WLAN access points to find the location of a mobile device connected to the WLAN to an accuracy of 1 to 2 m. The Ekahau solution enables location information without additional hardware (given enough wireless access points); it is a purely software solution. Bluetooth[2] is a short-range networking technology (within, say, 10 m, typically) enabling devices close enough to each other to connect with one another. What comes for free is location information if Bluetooth is used. If a device can detect the presence of another device, it must be within a proximity of several

---

[1] Ekahau Web site http://www.ekahau.com.
[2] The official site on Bluetooth is http://www.bluetooth.com.

meters. Hallberg and Nilsson (2002) describe a positioning system using multiple Bluetooth access points. The recent Ubisense[3] sensor technology provides location information for tagged people and objects to an accuracy of 15 cm. Also interesting is Microsoft's EasyLiving Project (Brumitt et al., 2000), which uses images from video cameras to determine the position of objects in a 3-D model. However, more people moving around can cause frequent occlusions for the vision, which makes tracking difficult.

The variety of positioning technologies and short-range networking technologies will surely continue to improve in their reliability and accuracy as time passes. An extensive review of location systems for ubiquitous computing is Hightower and Borriello (2001).

Where can we put sensors? Clearly, the question depends on the application and the type of sensor. Sensors can be embedded in the environment as part of the room or within a car, worn on people, and even placed within people. Sensors in the environment can be used to detect activities within that environment (even human movements) or the location of people within the environment.

Increasingly, sensors are being used in cars to automate functionality such as detecting rain and having the wipers turn on automatically, switching headlights on automatically when it is dark, generating warnings as part of advanced cruise control if two cars get too near each other while moving on the highway, and generating warnings if the system detects that the driver is not handling the vehicle well, for example, from drowsiness (Ayoob et al., 2003). Various forms of in-car telematics are also used, including diagnostics and automatic reporting of wear and tear of car parts. Not all cars have these functionalities, and some are merely experimental at this stage, but there is an increasing trend toward such use of sensors, leading to smart cars. Sensors can be worn on people for health monitoring or to detect emotional states (Picard, 1997) based on measurements of physiological factors such as heart rate and temperature.

Networks of sensors might be deployed for specific applications (Zhao and Guibas, 2004), such as warehouse inventory management (e.g., with the use of RFID tags), automotive applications including telematics and systems providing traffic congestion warnings and route information, building monitoring and control (e.g., of lighting and temperature conditions), environmental monitoring (e.g., chemical hazard warning system, tracking flood and extreme weather conditions, earthquake detection with seismic sensors, and wildlife habitat monitoring), military battlefield intelligence and security and surveillance, infrastructure protection (e.g., water distribution and power grids), and context-aware computing (e.g., intelligent homes and responsive environments). Such sensor networks, com-

---

[3] http://www.ubisense.net/.

prising a number of sensors such as the motes[4] scattered over a particular area, can be queried using SQL-like languages (e.g., a system called TinyDB (Madden, 2003)) and programmed to transmit information at a suitable rate or to sleep at certain times to conserve battery power. Sensor nodes include augmented general-purpose PCs linked to microphones and cameras, Berkeley motes, and smart dust. Berkeley motes can have more than a hundred kilobytes of program memory, several kilobytes of RAM, and several hundred kilobytes of nonvolatile storage in a range of sizes, and use radio frequency communication (with a rate of several tens of kilobytes per second). Smart dust[5] consists of tiny microelectromechanical sensors (MEMS) that can detect light intensity, temperature, humidity, and vibrations, and can transmit data wirelessly back to base stations. Such smart dust can be small enough to be scattered over an area and even suspended in air. They can be used in military applications to track enemy troops or to track patients in a hospital room. Sensors can also be programmed in the popular programming language of Java with the introduction of Sun's SPOT (Small Programmable Object Technology) wireless sensor devices, which hosts a Java 2 Micro Edition (J2ME)-compliant Java Virtual Machine (Smith et al., 2005).

Recent pervasive computing research has investigated the use of sensors to recognize everyday situations that humans or devices find themselves in. For example, in Fogarty et al. (2005), sensors are used to predict human interruptibility. The research showed that a microphone in the corner of an office, the time of day, a sensor to detect if a phone is in use, and information about mouse and keyboard activity can be used to estimate human interruptibility. Similarly, Ho and Intille (2005) looked at the perceived burden due to interruptions arising from mobile devices. They identified factors that influence a person's interruptibility at a given time, such as the current activity of the user, the utility of the message, the emotional state of the user, the time it takes to comprehend the interruption and act on it, previous and future activities, social engagement level, and social expectations. Focusing more on sensors carried or worn by users rather than those fixed in the environment, accelerometers were used to help recognize transitions in physical activity (e.g., from sitting to walking, walking to sitting, sitting to standing, and standing to sitting). The study showed that messages received during these physical activity transitions are more positively viewed than those messages delivered at random. In Gellersen et al. (2002), sensors (e.g., light sensors, accelero-

---

[4] A company that markets motes has the Web site http: ⁄ www.xbow.com/Products⁄ Wireless_Sensor_Networks.htm.

[5] See http://robotics.eecs.berkeley.edu/~pister/SmartDust/and http://www-bsac.eecs. berkeley.edu/archive/users/warneke-brett/SmartDust⁄.

meter) are attached to a mobile phone, which can determine the context of the phone, e.g., whether it is on the table, in the pocket, and so on. Specialized sets of sensors can be attached to everyday objects to add context-aware behaviors. The Smart-Its (Holmquist et al., 2004) is a stick-on computer with a set of sensors (i.e., light, sound, pressure, acceleration, and temperature) and the ability to communicate with other Smart-Its. Attached to everyday objects, they could lead to new applications such as load-sensing furniture and smart restaurants.

From the foregoing discussion, we see that sensors can be employed in everyday settings and in specialized applications. The ubiquitous computing wave encourages the thinking that sensors could proliferate in the environment and be worn somewhat unobtrusively (perhaps on normal apparel). The combination of information from different sensors provides a window into the real world for computer systems, but there exists a challenge in reasoning with sensor information and with the context information acquired to build a coherent and consistent picture of some part of the world. Also, there may be more than one way to recognize the same situation, using a different combination of sensors. What combination of sensors is best for acquiring context information to recognize a particular real-world situation is a question that application developers need to consider, where "best" might be in terms of cost, ease of deployment, and existing infrastructure. In Schmidt's Ph.D. dissertation (Schmidt, 2002), some constraints on sensing technologies used for context-aware devices include design and usability, energy consumption, calibration, start-up time, robustness and reliability, portability, size and weight, unobtrusiveness, social acceptance and user concern, costs, and precision and openness.

Sensors are almost as varied as their applications. According to Klein (1999), a sensor is a transducer (front-end hardware) that "converts the energy entering its aperture into lower frequencies from which target and background discrimination information is extracted in the data processor." One can broadly define sensor as any device, hardware or software, or their combination, that can be used to acquire context information. This definition of sensor is broad; devices not normally thought of as sensors might also be used to return context information and, therefore, are sensors under this definition, for example, the computer clock accessed using an operating system call or a video camera. Levels of abstraction of context information is a key idea related to our definition of sensor. A thermometer can provide temperature readings, but an application querying a Web service to return the current temperature can also be regarded as a sensor from the perspective of an application. Location information can be returned at different levels of abstraction. A "location sensor" might return sets of coordinates, or the equivalent (and more intelligible) name of a suburb, even in relative terms, describing the location as "where your mother lives." Compared to sensor technology work by Klein (1999), of rising interest in pervasive computing is context information presented at a higher level of abstraction. In addition, the kinds of situations that pervasive computing highlights, as compared to earlier military applications, are those which relate to everyday living, whether at home, in the office, or elsewhere.

It is often useful to hide (under some abstraction) how a type of context information is acquired. For example, there may be several methods for finding the location of a person (within a given range of accuracy), and all are equivalent in the sense that any of the methods can be used to return the same context information. We might term any such method the *location sensor*, whatever the actual mechanism. An application that uses such abstraction can work regardless of the actual underlying method employed, or the underlying method can be changed without modifying the application.

2.2.2 Thinking

In philosophical history, we have two schools of thought about the way knowledge is acquired: the rationalists and the empiricists. The rationalists attempt to gain knowledge about issues and the world (e.g., God, humanity, substance, space, etc.) by pure reasoning alone, whereas the empiricist attempts to gain such knowledge via experiences as perceived through the senses and stored in the memory. Without delving into philosophical discussions, a mix of the two or a compromise between these two extremes can be considered where some information is first perceived via the senses, and then reasoning is employed to infer more knowledge. Indeed, such a model underlies a general technique for building context-aware systems, which involves acquiring sensor information and then reasoning with some information to obtain knowledge; such knowledge together with other (perhaps built-in) knowledge can then be used to infer further knowledge, in particular, knowledge about the context or situation of entities. A sophisticated interleaving of sensing, reasoning, and acting might also be useful in an application.

Once data is obtained using a collection of sensors, the task is to utilize such data and to make sense of it. Based on our broad definition of sensors, the data can come in many different forms and can be discrete values or continuous series of values.

Earlier work on multisensor data fusion (Hall, 2001) has been used in fields such as robotics to estimate physical quantities such as angles and distances in the real world. Mathematical equations relate sensor observations with known quantities and variables whose values are to be discovered. Sensor fusion here involves taking a number of sensor observations from one or more sensors and solving these equations to find values for these variables. The focus here is on quantities and observations which can be captured mathematically, typically, using a set of linear equations.

With hardware sensors of the kind mentioned by Klein (1999), various techniques have been used to make sense of the sensor data for detection, classification, and identification of objects using sensors, including the following:

- Physical mathematical models such as Kalman filtering
- Feature-based inference techniques, such as cluster algorithms, correlation measures, pattern recognition, neural networks, and Dempster–Shafer and Bayesian reasoning to deal with uncertainties
- Cognitive-based models such as logical templates, knowledge bases, and fuzzy logic

With our more abstract definition of sensors, many such techniques have also been used to reason with context information, with pervasive computing applications in mind.

Recent work has utilized sensors to observe patterns in real life. For example, in a study by Clarkson (2002), data was collected using two cameras, a microphone, and an orientation sensor worn by a user for 100 days. Scene segmentation and quantitative analysis techniques were then applied on the data to extract patterns in daily life, demonstrating that the life of a person can be segmented into distinct situations distinguished by location and activity at that location and that a "person's life is not an ever-expanding list of unique situations" but that there is a great deal of regularity in daily situations.

Apart from data analysis techniques, knowledge-based approaches have been used to represent and manipulate context information acquired from sensors. Such approaches attempt a rich, explicit model of context to facilitate more sophisticated reasoning. Examples of rich models of context include location models such as in Bauer et al. (2002), augmented-world models (Nicklas and Mitschang, 2001), and activity models (e.g., see Bardram (2005)).

Location models can be used in conjunction with positioning technologies mentioned earlier to provide location information with more structure. According to Bauer et al. (2002), location models can be geometric (with respect to a 2-D or 3-D coordinate system); symbolic such as building, floor, and room numbers; and hybrid (i.e., specified using combinations of geometric and symbolic information). One could imagine mapping geometric coordinates into more descriptive symbolic labels.

A lattice-based symbolic model and a graph-based symbolic model are mentioned in Bauer et al. (2002). In the lattice-based location model, the spatial containment relationship is used to impose a partial order on locations, reflecting a hierarchical view of spaces. In the graph-based location model, edges labeled with distance measures connect nodes representing locations. Where location is coupled with time, spatial-temporal models might be devised. The actual location model used would depend on application requirements. In Loke et al. (2006), a hierarchical model of spaces within a shopping center (called Chadstone in Melbourne, Australia) was used as a means to index mobile services. For example, particular stores are located within a large departmental store, which, in turn, is within the shopping center itself; similarly, a retail outlet is located within a particular floor, which, in turn, is within the shopping center.

The Nexus spatial world model (Nicklas and Mitschang, 2001) aims at creating a comprehensive spatial world model with stationary and mobile objects, integrating virtual objects and representations of real-world objects to form the augmented-world model. The Augmented World Modeling Language (AWML) is used to describe objects in the augmented world, including object geometry, coordinate systems, symbolic descriptions, and relationships between objects. The corresponding Augmented World Query Language (AWQL) can be used to query descriptions in AWML.

Apart from location, semantic proximity is another interesting physical relationship that can be perceived and exploited. Antifakos and Schiele (2002) identify three levels of semantic proximity:

1. Different locations having similar static states of the environment (e.g., similar weather) are considered semantically close
2. Different devices that detect similar and simultaneous dynamics in the environment (e.g., detect similar light changes and sounds)
3. Different devices attached to the same object or moving together

Two devices in the same or different locations might share such semantic closeness.

Updating of the aforementioned world models, especially when they mirror real-world objects that can move, can be a difficult task, and there is an issue to manage such updates efficiently and to perhaps support approximate answers to queries.

In Harle and Hopper (2003), a 3-D world model is created using an ultrasonic positioning system via ray-tracing techniques, in which ultrasonic signals (rays) are sent from a transmitter to receivers. Such models can be updated easily to detect appearance, disappearance, and movement of objects. It is such technology based on Ultrawideband radio which is used for Ubisense mentioned earlier.

Activity models have also been employed to model the notion of activity of a person or a group as a context attribute. According to Bardram (2005), the basis of activity-based computing is that "user activities become first-class entities that are represented explicitly, and activities are inherently collaborative, treating single-user activities as collaborative activities that just happen to have only one participant." The notion of activity becomes a primary modeling concept, wherein different applications or services can be grouped with the activity (such as "prescribing medicine for John Smith") as a focus.

As mentioned in Chapter 1, ontologies have also been used to provide a comprehensive model of different types of context information which can be used to describe situations for particular domains. Once context information is represented, it can be reasoned about in a logical way and different entities can understand and utilize the knowledge. Uses of such ontologies include (Ranganathan et al., 2004) configuration management (allowing different components to automatically discover and collaborate with other components), semantic discovery and matching of entities, easier interoperability between entities, and improvement of the robustness and portability of context-aware applications. Chen et al. (2004) have developed an ontology to represent concepts such as place, room, building, speaker, and person, which can be employed to describe context. Other concepts such as "Meeting," "PersonInBuilding," and "MeetingParticipant" refer to situations. One can also write rules describing conditions (stated in terms of the context of entities), under which someone could be in a meeting or someone could be the speaker, and so on. A similar ontology is given in Gu et al., (2004a) but with notable differences for modeling the quality of context information (such as its accuracy, resolution, and freshness). Logic rules are also used to describe conditions under which a situation would be deemed to be occurring. The ontology in Matheus et al. (2003) employs concepts from situation theory in which situations are viewed as collections of propositions (in the form of relations between objects).

Studies such as Ranganathan and Campbell (2003) have applied first-order logic-based formalisms to represent context and situations, and rules that map situations to required actions. Prolog is the logic programming language used, providing executable rules for practical applications. Prolog rules relate context information (conditions in a rule) to situations (conclusion in the rule). The approach in Loke (2005) introduces the additional abstraction of situation programs for representing situations and model sensors as special kinds of predicates.

Other context representation formalisms include contextual graphs (Brezillon, 2003). Padovitz et al. (2004) provide a spatial view of context, introducing the notion of context spaces. Context attributes form the axes of a multidimensional space, situations are represented by regions in the space, and the current set of sensor readings (i.e., a tuple of values for the context attributes) are represented by a point in the space.

There is a diversity of approaches used as already seen, and there remains an active area of research in finding the right level of abstraction in which to represent situations and context for different applications.

One very important dimension to reasoning with context information is dealing with uncertainty. It is not surprising that existing formalisms for representing different kinds of uncertainty, including probability value assignments, and degrees of set membership for vagueness have been used; e.g., Bayesian reasoning (Gu et al., 2004b), Dempster–Shafer techniques (Wu, 2003), and fuzzy logic (Byun and Cheverst, 2003).

Dey and Mankoff (2005) suggest involving the user, or *mediation*, referring to the dialogue between user and system in resolving ambiguity, which cannot be handled well even with AI techniques. Principles for doing so include providing multiple redundant mediation techniques, use of defaults to minimize user intervention, and retaining ambiguity until required to be resolved.

2.2.3 Acting

Once context information has been gathered or situations recognized, actions are taken. Effectors and the actions to be taken are application specific, and the action itself might be to perform further sensing. Performance is a consideration. Actions might need to be performed in time for it to be of use to the users, and before the situation which triggered the action changes. Another consideration is control. Ideally, the user should retain control and be able to override actions, cancel actions, stop actions, or reverse the effect of actions. However, only some of these are possible, depending on the nature of the action. Agent-based systems might also employ planning before acting, so complex actions comprising sequences of actions are taken, perhaps interleaved with further sensing and reasoning.

2.3 AN ABSTRACT ARCHITECTURE

Design considerations for building context-aware systems relate to the previously mentioned three phases of sensing, thinking, and acting. One has to consider the situations to be recognized and the context information that would be available and can be feasibly acquired, including sensors to be used. The appropriate reasoning technique is then chosen, ranging from simple event-condition rules to sophisticated AI techniques. Some data or knowledge extracted via reasoning might also be stored. Finally,

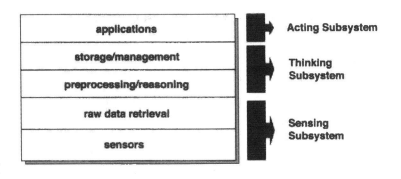

Figure 2.1   Abstract layered architecture for context-aware systems.

appropriate effectors, hardware, and software are employed. The distribution of each of these components would depend on the application, and all these components might be situated on the same machine or in a distributed infrastructure. The sensing subsystem, the thinking subsystem, and the acting subsystem need to be connected.

Several general architectures are given in the literature for context-aware systems. Figure 2.1 shows an abstract layered architecture from Baldauf and Dustdar (2004), labeled with subsystem divisions.

The word "sensors" at the bottom layer relates to the definition of sensors used for raw data retrieval, given earlier. Subsequent preprocessing or reasoning of the data is then carried out, and resulting context information stored. Storage and management of context can be sophisticated with support for querying, further reasoning, and updates, or much simpler, depending on application requirements. The three subsystems of acting, thinking, and sensing involve sensors and raw data retrieval, preprocessing and management of context, and application-dependent actions, respectively.

Each subsystem might be complex and decoupled from each other, or tightly integrated into one device. Each subsystem itself might be a collection of distributed components. An application might have its own thinking and sensing subsystem, rather than use a shared reasoning and sensing subsystem as part of an infrastructure (as we will see later). Between subsystems, there might be generic interfaces (perhaps based on the general notion of services), which allow each subsystem to interact with another subsystem without knowing the underlying details. Such abstraction facilitates change in one layer without affecting others. For example, reasoning algorithms might be updated, but such reasoning services being invoked via interface calls do not require changes on the caller side.

2.4 INFRASTRUCTURES, MIDDLEWARE, AND TOOLKITS

From an abstract viewpoint, frameworks such as SOCAM (Service-Oriented Context-Aware Middleware), Gaia, CoBrA (Context Broker Architecture), the Context Toolkit, the Sentient Object Model, and Hydrogen follow the general structure as that shown earlier, but aim to separate applications clearly from context-aware functionality.

According to Hong and Landay (2001), an infrastructure is "a well-established, pervasive, reliable, and publicly accessible set of technologies that acts as a foundation for other systems." The idea is that such an infrastructure can greatly simplify the development of context-aware applications by providing a common set of functionality, which applications can simply utilize. Indeed, the layered architecture described earlier suggests subsystems that can be abstracted into an infrastructure, whether the sensing subsystem, the thinking subsystem, or both.

Advantages of the infrastructure approach, according to Hong and Landay, are as follows:

■ Independence from hardware, operating system, and programming language: Standard data formats and protocols can be used to access the services of the infrastructure, thereby allowing heterogeneous clients as long as they adhere to the standards.

■ Improved capabilities for maintenance and evolution: Changes including improvements and fixes to the components of the subsystems can be made independently of applications, just as subsystems can be modified independently of other subsystems.

■ Sharing of sensors, processing power, and services: Resources can be shared by multiple applications via an infrastructure, thereby simplifying not only application development but also deployment and facilitating interoperability among applications. Also, applications might not in themselves have the required capabilities or computational resources for reasoning and storage of context, and applications might not be equipped with their own sensors, and so an infrastructure enables such applications to have context-aware capabilities easily.

Many such context-aware frameworks have emerged in recent years, including those mentioned earlier, as well as Henricksen and Indulska's (2004) framework using situation predicates in first-order logic and condition-action triggers, Context Fabric,[6] and Gu et al.'s (2005) service-oriented framework, each with its particular emphasis being investigated

---

[6] http://guir.berkeley.edu/projects/confab/.

in the project. In a survey of ubiquitous computing frameworks, Endres et al. (2005) studied projects mainly in Europe and the United States.

One could also consider multi-infrastructure approaches when an application might utilize the capabilities of different support infrastructures at different times (e.g., when the application is moved from one location to another location supported by a different infrastructure).

2.5 ISSUES OF SECURITY, PRIVACY, AND EFFICIENCY

There have been discussions concerning the privacy and security of context information (e.g., location). Because context information such as location can be abused if it passes to the wrong hands, such information must be managed within secure boundaries. Hong's Ph.D. thesis (2005) proposed Confab, a toolkit for managing personal context information, which helps end users make informed decisions about how and where their personal information is utilized, including whom such information should be exposed to and for how long. Hence, context information about people may be gathered in a way that is unobtrusive to users, but users might need to help manage such information or approve of their uses.

Context information must be acquired in an efficient manner (Satyanarayanan, 2001). Otherwise, it is possible that the world would have changed before action could be taken based on the acquired context information. For example, suppose we have an application that disseminates advertisements over a wireless network to a user's mobile device, depending on the user's location, while the user is on the move. Then, if the user is moving fast enough (or if the system is too slow in finding and utilizing the user's location information), it is possible that the advertisements are not received in time and so corresponds incorrectly to the location that the user was last in. In the Mobile Hanging Services system, services (in the form of mobile code) are delivered to users' mobile devices, depending on the user's location. Because it takes some time for the user's location to be obtained and for the code to be downloaded, the user encounters an uncomfortable wait when at a new location unless the system is adequately efficient or somehow prefetches the code for a location before the user arrives at the location. Chapter 3 discusses this further.

2.6 SUMMARY

We have reviewed the abstract architecture and basic elements of context-aware systems. We have also noted technologies and techniques utilized in sensing, thinking, and acting subsystems. The following chapters will consider architectures as developed in applications and scenarios, with the abstract architecture as an underlying blueprint.

REFERENCES

Andersson, C., *GPRS and 3G Wireless Applications: Professional Developer's Guide*, John Wiley & Sons, U.S.A., 2001.

Antifakos, S. and Schiele, B., Beyond position awareness, *Personal and Ubiquitous Computing* 6(5/6), 313–317, 2002, Springer.

Ayoob, E.M., Grace, R., and Steinfeld, A., A user-centered drowsy-driver detection and warning system, *Proceedings of the Conference on Designing for User Experiences*, ACM Press, 2003, pp. 1–4.

Bardram, J.E., Activity-based computing: support for mobility and collaboration in ubiquitous computing, *Personal and Ubiquitous Computing* 9, 312–322, 2005, Springer.

Bauer, M., Becker, C., and Rothermel, K., Location models from the perspective of context-aware applications and mobile ad hoc networks, *Personal and Ubiquitous Computing* 6(5), 322–328, December, 2002, Springer.

Baldauf, M. and Dustdar, S., A Survey on Context-Aware Systems. Technical Report Number TUV-1841-2004-24, November 2004, available at http://www.infosys.tuwien.ac.at/Staff/sd/papers/TUV-1841-2004-24.pdf.

Becker, C. and Durr, F., On location models for ubiquitous computing, *Personal and Ubiquitous Computing* 9, 20–31, 2005, Springer.

Brezillon, P., Representation of procedures and practices in contextual graphs, *The Knowledge Engineering Review* 18(2), 147–174, 2003, Cambridge University Press.

Brumitt, B., Meyers, B., Krumm, J., Kern, A., and Shafer, S.A., Easyliving: technologies for intelligent environments, *Proceedings of HUC*, 2000, pp. 12–29.

Byun, H.E. and Cheverst, K., Supporting proactive "intelligent" behaviour: the problem of uncertainty, *Proceedings of the Workshop on User Modelling for Ubiquitous Computing*, at the 9th International Conference on User Modelling, Pittsburg, PA, June 2003, available at http://www.di.uniba.it/~ubium03/byun-8.pdf.

Chen, G. and Kotz, D., A Survey of Context-Aware Mobile Computing Research, Technical Report TR2000-381, Department of Computer Science, Dartmouth College, November 2000.

Chen, H., Finin, T., and Joshi, A., An ontology for context-aware pervasive computing environments, *The Knowledge Engineering Review (special issue on Ontologies for Distributed Systems)* 18(3), 197–207, 2004.

Clarkson, B.P., Life Patterns: Structure from Wearable Sensors, Ph.D. thesis, Massachusetts Institute of Technology, 2002.

Dey, A. and Mankoff, J., Designing mediation for context-aware applications, *ACM Transactions on Computer Human Interaction (TOCHI)*, 12(1), 53–80, March 2005.

Endres, C., Butz, A., and MacWilliams, A., A survey of software infrastructures and frameworks for ubiquitous computing, *Mobile Information Systems Journal* 1(1), January–March 2005, IOS Press.

Etzioni, O., Moving up the information food chain: Deploying softbots on the World Wide Web, *AI Magazine* 18(2), 1997, pp. 11–18.

Fogarty, J., Hudson, S.E, Atkeson, C.G., Avrahami, D., Forlizzi, J., Kiesler, S., Lee, J.C., and Yang, J., Predicting human interruptibility with sensors, *ACM Transactions on Computer-Human Interaction (TOCHI)* 12(1), 119–146, March 2005.

Gellersen, H.-W., Schmidt, A., and Beigl, M., Multi-sensor context-awareness in mobile devices and smart artifacts, *Mobile Networks and Applications (MONET)*, 7(5), October 2002, pp. 341–351, Kluwer Academic Publishers.

Gu, T., Wang, X.H., Pung, H.K., and Zhang, D.Q., An Ontology-based Context Model in Intelligent Environments. Proceedings of Communication Networks and Distributed Systems Modeling and Simulation Conference (CNDS 2004), pp. 270-275, San Diego, CA, January 2004a.

Gu, T., Pung, H.K., and Zhang, D.Q., A Bayesian approach for dealing with uncertain contexts, *Proceedings of the Second International Conference on Pervasive Computing (Pervasive 2004)*, in the book *Advances in Pervasive Computing* published by the Austrian Computer Society, Vol. 176, Vienna, Austria, April 2004b.

Gu, T., Pung, H.K., and Zhang, D.Q., A service-oriented middleware for building context-aware services, *Journal of Network and Computer Applications (JNCA)* 28(1), 1–18, January 2005, Elsevier.

Hall, D., *Handbook of Multisensor Data Fusion*, CRC Press, Boca Raton, FL, 2001.

Hallberg, J. and Nilsson, M., Positioning with Bluetooth, IrDA and RFID, Master's thesis, 2002, available at http://www.cdt.luth.se/~amaino/papers/Positioning_with_Bluetooth_IrDA_and_RFID.pdf.

Harle, R.K. and Hopper, A., Building world models by ray-tracing within ceiling-mounted positioning systems, *Proceedings of Ubicomp*, 2003, Springer-Verlag, Germany, pp. 1–17.

Henricksen, K. and Indulska, J., A software engineering framework for context-aware pervasive computing, *Proceedings of the 2nd IEEE International Conference on Pervasive Computing and Communications (PerCom)*, March 2004, IEEE Computer Society, U.S.A., pp. 77–86.

Hightower, J. and Borriello, G., Location systems for ubiquitous computing, *IEEE Computer* 34(8), 57–66, 2001, IEEE Computer Society.

Ho, J. and Intille, S., Using context-aware computing to reduce the perceived burden of interruptions from mobile devices, *Proceedings of CHI 2005 Connect: Conference on Human Factors in Computing Systems*, ACM Press, 2005, pp. 909–918.

Holmquist, L.E., Gellersen, H.-W., Kortuem, G., Schmidt, A., Strohbach, M., Antifakos, S., Michahelles, F., Schiele, B., Beigl, M., and Maze, R., Building intelligent environments with smart-its, *IEEE Computer Graphics and Applications*, 24(1), 56–64, 2004.

Hong, J.I., An Architecture for Privacy-Sensitive Ubiquitous Computing, Unpublished Ph.D. thesis, University of California at Berkeley, Computer Science Division, Berkeley, 2005.

Hong, J.I. and Landay, J.A., An infrastructure approach to context-aware computing, *Human-Computer Interaction* 16(2, 3, 4), 287–303, 2001.

Jagoe, A., *Mobile Location Services: The Definitive Guide*, Pearson Education, U.S.A., 2002.

Klein, L.A., *Sensor and Data Fusion Concepts and Applications*, 2nd ed., Society of Photo-Optical Instrumentation Engineers (SPIE), U.S.A., 1999.

Loke, S.W., Representing and reasoning with situations for context-aware pervasive computing: a logic programming perspective, *Knowledge Engineering Review*, 19(3), 213–233, 2005, Cambridge University Press.

Loke, S.W., Tam, O.K., Krishnaswamy, S., and Cuce, S., Bringing order to mobile services: a weighted approach for prioritizing ambient services, submitted to the *Journal of Pervasive Computing and Communications*, 2006.

Lahiri, S., *RFID Sourcebook*, IBM Press, U.S.A., 2005.

Madden, S., The Design and Evaluation of a Query Processing Architecture for Sensor Networks, UC Berkeley, Ph.D. thesis, 2003.

Matheus, C., Kokar, M., and Baclawski, K.., A core ontology for situation awareness, in *Proceedings of FUSION*, July 2003, pp. 545–552, available at http://vistology.com/papers/FUSION03.pdf.

Mitchell, K., A Survey of Context-Aware Computing, Report, Lancaster University, March 2002, available at http://www.comp.lancs.ac.uk/~km/papers/ContextAwarenessSurvey.pdf.

Nicklas, D. and Mitschang, B., The NEXUS augmented world model: an extensible approach for mobile, spatially-aware applications, *Proceedings of the 7th International Conference on Object-Oriented Information Systems*, Calgary, 2001.

Padovitz, A., Loke, S.W., and Zaslavsky, A., Towards a theory of context spaces, *Proceedings of the Workshop on Context Modelling and Reasoning (COMOREA)*, PerCom Workshops, IEEE Computer Society, U.S.A., 2004, pp. 38–42.

Picard, R., *Affective Computing*, MIT Press, 1997.

Ranganathan, A. and Campbell, R.H., An infrastructure for context-awareness based on first order logic, *Personal and Ubiquitous Computing Journal* 7, 353–364, 2003.

Ranganathan, A., McGrath, R.E., Campbell, R.H., and Mickunas, M.D., Use of Ontologies in a Pervasive Computing Environment, *The Knowledge Engineering Review* 18(3), 209–220, 2004, Cambridge University Press.

Satyanarayanan, M., Pervasive computing: vision and challenges, *IEEE Personal Communications* 8, 10–17, August 2001.

Schmidt, A., Ubiquitous Computing — Computing in Context, Ph.D. thesis, Computing Department, Lancaster University, 2002.

Smith, R.B., Cifuentes, C., and Simon, D. Enabling Java for Small Wireless Devices with Squawk and SpotWorld, *Proceedings of the Workshop on Building Software for Pervasive Computing at OOPSLA*, 2005, available at http://www.ics.uci.edu/~lopes/bspc05/papers/smith.pdf.

Want, R., Fishkin, K.P., Gujar, A., and Harrison, B.L., Bridging physical and virtual worlds with electronic tags, *Proceedings of the Conference on Computer Human Interaction (CHI)*, 1999, pp. 370–377.

Wu, H., Sensor Data Fusion for Context-Aware Computing Using Dempster-Shafer Theory, Ph.D. thesis, The Robotics Institute, Carnegie Mellon University, 2003.

Zhao, F. and Guibas, L., *Wireless Sensor Networks: An Information Processing Approach*, Morgan Kaufmann, U.S.A., 2004.

3

CONTEXT-AWARE
MOBILE SERVICES

Service-oriented computing has the notion of services as its central operative idea, made popular by developments in Web service technologies. Computationally, services are an abstraction of a unit of functionality, whereas to end users it conveys the idea of applications that can be called upon as needed to perform, if not aid, users' tasks. Mobile computing implies the tendency for users to change their situation often (at least their location, for instance), and such changes can be exploited by the system to proactively tailor services or to present services according to the user's current situation (e.g., location-based services and location-based E-commerce applications); i.e., the services are said to be effectively context aware or they behave in a way that demonstrates awareness of the user's current context.

This chapter first briefly reviews the development of mobile services, the context useful for mobile services, and some examples of applications. Then, the chapter puts forth ideas building up on mobile service infrastructure, including those of ambient services, mobile electronic communities (building on location-based services), and technical enhancements related to advanced delivery of mobile services (e.g., use of mobile code, and multiagent systems for location-based auctions). Many of these ideas are exploratory, aimed at suggesting possibilities with context-aware mobile services.

3.1 THE RISE OF MOBILE SERVICES

From E-services available over the Internet supported by engines such as E-speak (Karp, 2003), integration with Web philosophy yields the notion of *Web services*. Singh and Huhns (2005) enumerated a list of definitions for Web service and arrived at "functionality that can be engaged over the Web." Engagement can mean invocation of a service as in invocation of a method in the programmatic sense to more sophisticated interactions over time. Web service comprises technology (XML-based formats and protocols) for describing such functionality and for transferring data to and from such services. Web services provide a way for a company to expose business functionality over the Internet and allow services sharing within the intranet of a business, with extranets, or between business partners.

With the development of the mobile Internet or mobile Web, such services can then be engaged over wireless networking technologies anywhere, anytime as long as the infrastructure supports the services. Wireless Application Protocol (Singhal et al., 2001) standards and applications were developed to support lightweight content that could fit within small resource-constrained devices and be controlled with the phone keypad. Such content, even if more efficiently processed with less memory and computational power, was poorer in multimedia functionality and expressiveness compared to the desktop Web. With the recent development of wide area high bandwidth wireless networking technologies such as 3G and 4G, multimedia content could be delivered to smartphones at high speeds, even allowing real-time videos of phone users to be exchanged. Such technologies and newer handsets enable multimedia services to be delivered to users anywhere (provided there is network coverage), anytime.

Location-based E-services became a popular idea for M-commerce and L-commerce (L for location). Examples of such services include finding the nearest ATM machine and making reservations with restaurants in the vicinity. Such services come in the form of queries whose results depend on the current location of the device from which the query is issued. Positioning technologies employed include GPS-based technologies. Web services accessed from mobile devices such as smartphones and personal digital assistants (PDAs) can be adapted or tailored according to such context information of the devices (and the user). Besides location, other information about the user or device can be used to adapt services or content delivered to services. Content adaptation becomes a key idea in ensuring that a mobile device receives information in a form it can manage,[1] appropriate to the device's available memory, computational power, and device interface capabilities. Other ways in which context might be used includes tailoring advertisements sent to mobile devices

---

[1] See IBM's WebSphere Everyplace at http://www-306.ibm.com/software/integration/wmqe/.

according to what is near the user (or the device the user is carrying),[2] navigational services, object finding, localized Internet search, and location-based auctions, as we will discuss later. Context changes can be used as a means to trigger changes of behavior of services.

3.2 CONTEXT FOR MOBILE DEVICE USERS

What information might be context in this category of applications? With mobility, location is a key context that changes often. The granularity of location becomes important depending on the application. Location can be viewed as a point in space or an area with comprehensible geographic boundary. Time and current activity of the user carrying the device, proximity to other objects and people, intentions of users, and data from sensors in the environment might also be exploited in applications. In Stolze and Strobel (2001), shopping roles such as "father shopping for a birthday present for his daughter" is used as context to select appropriate services for display to a shopper or to adapt the behavior of particular services.

3.3 LOCATION-BASED SERVICES

Applications can be indoor, outdoor, or supported both within buildings and outdoors. Important considerations are the positioning technology available and the network coverage, as well as device support for networking technologies. For example, it is possible for indoor services to be supported for a mobile device by a short-range wireless networking technology such as Bluetooth or infrared beacons, and then on moving indoors to outdoors, the device switches automatically to an outdoor networking technology and a corresponding outdoor positioning technology (e.g., based on wireless local area network [WLAN] or global positioning system [GPS], and accesses services via this network.

We can consider outdoor location-based applications, in which the position of a device (and its carrier) is determined by technologies such as GPS and cellular networks. Examples of such applications include navigation and real-time traffic, emergency services, travel guides, mobile yellow pages, location-based marketing, asset (e.g., vehicle fleet or packages) tracking, and theft control. The positioning technology determines the location of the object as found with respect to a digital map. Jagoe (2002) gives a comprehensive infrastructure for location-based services, including components for positioning, payment for services, authentica-

---

[2] As an example, see the European project on location-based advertising at http://www.e-lba.com/.

tion, and billing logic. The maps used depend on the areas over which locations are to be considered, and granularity of location information depends on the application (e.g., cells in the case of cellular networks).

3.4 AMBIENT SERVICES

Related to location as context for adapting services, it is possible to consider ambients, i.e., geographic boundaries of areas, as one type of such contextual information. By *ambient services*, we have in view services that are related to the surrounding physical environment of the user and are locally useful (i.e., they may not be relevant or useful beyond the boundaries of an area around the user). Compared to location-based services in general, we use the term ambient services to emphasize the association of services with logical areas with boundaries, so that it is possible to talk about these areas overlapping or one area being contained within another, and the user crossing such boundaries.

Such ambient services are naturally suggested by short-range wireless networking technologies (Leeper, 2001), such as WLANs (e.g., the IEEE 802.11b [or Wi-Fi] LANs) with access points having a limited range of roughly 100 m) and wireless personal area networks (WPANs) (e.g., Bluetooth [http://www.bluetooth.com] with a limited range of roughly 10 m). WLANs (or "hot-spots") are becoming ubiquitous and are starting to appear in homes, offices, and public places such as shopping complexes, airports, hotels, parks, and restaurants.[3] The Bryant Park Hotel in New York lets its staff service guests using information provided via wirelessly networked devices.[4] Starbucks cafe is providing WLAN services to customers. Also, Bluetooth public access points are emerging. Ericsson's Bluetooth Local Infotainment Point (BlipNet)[5] technology provides access points to information for Bluetooth-enabled PDAs and cell phones, and has undergone trials in public areas such as cafes, railways, and gas stations. The vision of the ubiquitous short-range networks is also articulated in the Point Servers concept.[6]

More recent efforts will enable roaming across different WLANs and Bluetooth networks, and even WLANs and Global System for Mobile Communications (GSM) networks.[7] For example, one could connect to a WLAN within range of its access point, hand off to a lower speed wireless telecommunications service once out of range, and then reconnect to

---

[3] A list of 802.11 access points can be found at http://www.80211hotspots.com/.

[4] See http://www.mobileinfo.com/News_2001/Issue34/Symbol_Bryant.htm.

[5] http://www.blipsystems.com/.

[6] http://www.pointservers.org/.

[7] See http://www.cisco.com/en/US/products/ps5940/products_white_paper0900aecd80374a00.shtml.

another access point in a different WLAN. An individual can move out of one cafe (and so out of that cafe's network) and into another cafe (and so into this cafe's network) while still in some other network. It is also possible for a user to be within the range of and connected to multiple networks at the same time, to some networks containing others, or contained in or overlapping with others, provided the user's device has that capability or the different devices of the user (themselves interconnected by a WPAN) are each connected to a different network. For example, one could be in a cafe in a shopping complex in some town and so could then be within and connected to three networks at the same time: the café's WLAN, the shopping complex's WLAN, and the town's network (and, perhaps, even to a fourth one — a wide area mobile Internet).

What granularity of boundaries or logical areas to use for ambient services would depend on the application. A logical area will have certain ambient services associated with it, which might or might not also be associated with ambient services in another logical area. Figure 3.1 shows the boundaries for services, with logical areas P, Q, R, and S (Loke and Zaslavsky, 2004).

In Figure 3.1, we assume that a service p's boundary is P; i.e., only while a user U is in P is the service p active for U. Similarly, Q is the boundary for a service q, R is the boundary for a service r, and S is the boundary for a service s. The dark spot denotes a person located within the service boundaries P, Q, and R and therefore able to utilize services p, q, and r, but not able to utilize service s, not being within the boundary S of s. Such service boundaries may be located within another, for example, Q within P. Q could represent a service to obtain information about a

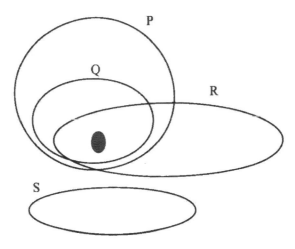

Figure 3.1   Boundaries P, Q, R, and S for services.

particular museum exhibit, and P might be the boundary for the services of the museum, and only when the user is 1 m away from the exhibit is the user within Q. The developer of a service can use such service boundaries to demarcate geographically the area of relevance and utility of a service. Such boundaries also provide a filtering criteria for the user who then needs only to be informed of services that he or she is currently within the boundaries of. We admit that not all services will have a meaningful geographic area of relevance but assume that ambient services by definition do. The notion of service boundaries not only conveys information about the application characteristics of the services themselves but can also be used to capture the administrative domains of services; service providers can structure their services into such logical areas.

We can have services with relatively fine scope such as an office room, a lecture theatre, or a building floor. Logical areas of relevance for services of much larger granularity can be considered on such scale as a shopping mall, a street, a suburb, or a town, and relevance areas of smaller granularity can be considered, such as a space 0.5 m from a museum exhibit, or a telephone booth. A set of services is associated with each such logical area (e.g., specific library services while within a library, information services on specific museum exhibits of a museum hall, and accessing particular printers when in some part of an office). When an individual is within a logical area, the services associated with the area can be invoked by the user.

Because of geographic overlaps and nesting, a user can be in several such logical areas at the same time, and as the user enters one or more logical areas, moving beyond the boundaries of some services and moving into others, his or her ambient services can be automatically discovered and enabled, and updated on the user's device. When a user is in several logical areas at the same time, an important issue is how to decide what combination of services should be enabled for a user at a given time. This combination of services may take a number of different forms. For example, the same or similar services may be combined, or services in one area may have precedence over similar services in another area.

More precisely, the scope D of a set of services for a user U is the geographic area in which the services S' for the user U is available such that $S' \subseteq S$, where S is the set of all the services available in D. So, for example, in our model, let Di and Dj be two logical areas where $i \neq j$, and let Sa and Sb be two sets of services where $a \neq b$. If Sa and Sb are the sets of services associated with areas Di and Dj, respectively, and if the user is within Di and Dj at this time, then a simple case is that the ambient services available to the user is $Sa \cup Sb$ (set union of Sa and Sb). So, in this example, for a given user U and areas Di and Dj, a set of services $Sa \cup Sb$ comprises all the ambient services of the user U (assuming

Di and *Dj* are mutually exclusive). But this might not be the case; for example, we might have a case in which the user is within *Di* and *Dj*, but he or she is only allowed to use the services of *Di*. This may be due to the constraints imposed by the service provider of the area or to the user's own preference of what services he or she wants enabled at a particular area. Thus, the services available to a user at a particular area are not all the services associated with the areas containing that area but, in some cases, only a combination of some of the services associated with the areas.

The concepts discussed are applicable to both outdoor and indoor logical areas. We now consider natural boundaries in shopping centers and office buildings to illustrate indoor logical areas.

3.4.1 PointRock Example

We consider a hypothetical shopping center called PointRock, located in Melbourne, Australia. PointRock Shopping Centre is a complex shopping mall that comprises many stores, shops, and services, such as ANK Bank and post offices. Bazar is a store with many smaller retailers inside — for example, Jesprit-branded clothing. Once users enter a predefined logical area, the associated services will be available to users for selection. For example, consider Sally currently shopping in Jesprit; she is located in the predefined logical area: *Jesprit logical area*. Thus, all the ambient services associated with the Jesprit logical area will be available to Sally. Such services could, for example, include an informational service about Jesprit clothing or a price check facility. In the same way in which information about a museum exhibit can be beamed to a visitor's mobile device (e.g., via an infrared or Bluetooth connection or an RFID reader in the device, the museum object being RFID-tagged), information about items of Jesprit clothing can be acquired.

Figure 3.2 shows a simplified floor plan of a shopping center, inspired by an actual shopping center in Melbourne, Chadstone.[8] We observe the following logical areas: James Jones, Tarju, and Bazar are three departmental stores, and Shop-Low and Toles are two supermarkets. ANK Bank's logical area is shown; Jesprit is a section within Bazar for display and retail of Jesprit-branded clothing; the boxes indicate subdivisions of the space for retail shops. One could develop ambient services pertinent to each such logical area.

There could be services whose scope is the entire PointRock (useful and accessible while the user is in PointRock) or services useful only while in, say, James Jones or Bazar. There could also be services specific to the

---

[8] http://www.chadstoneshopping.com.au.

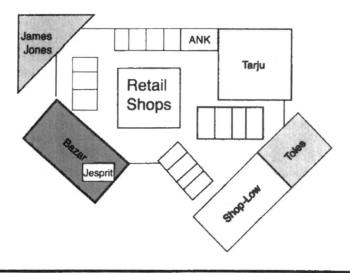

Figure 3.2 A sketch of logical areas in a shopping mall.

logical area of "supermarkets" (which is a combination of the areas of Shop-Low and Toles). For example, a service to guide the user to the nearest restaurant, café, or restroom can be PointRock-wide, whereas a service to guide a user to the specific shelf on which a type of clothing can be found might only be applicable in Bazar. Similarly, a service to guide the user to the particular shelf on which a certain type of can food can be found might have the scope of supermarkets. There could be a parking lot search service while users, in cars, are in the parking space of PointRock.

3.4.2   Future E-Martketplaces

As noted by Loke (2003), we can envision E-marketplaces as counterparts of physical marketplaces; we have the scenario of one in a physical marketplace down the street (e.g., a shopping mall) able to access from our devices not only some global E-marketplace or store (e.g., Amazon.com) but also an E-marketplace that is locally relevant and enhances the experience in the physical marketplace itself. Such location-based E-marketplaces can transcend geographic features spanning not necessarily continents but neighborhoods, streets, buildings, or shops. We might have a virtual sports shopping mall, which is an E-marketplace representing all sports shops within a 2-km radius, or a virtual designer-jeans shopping mall representing shops selling such items along a famous street. Hence, conceptually, we have a superimposing of E-marketplaces and physical marketplaces, with added value to vendors and potential customers. For example, there is opportunity to extend the boundaries of stores beyond the physical store space — proximity advertising can act as a kind of "tractor beam" to attract prospective customers to visit the shop (Newell and Newell, 2001); people can turn their "sale radar" on (opting On) to be a prime candidate for location-based targeted marketing communications, reply to ads to preregister or hold an item or preorder for instant consumption, or get into a "virtual red carpet" shopping tour. This means that the "conceptual size" of shops increases, and there can be multiple "conceptual organization" of shops. There is also opportunity for any-time–anywhere B2B collaboration; for example, several stores at nearby locations can combine resources to create instant packages to lure buyers, or several businesses can provide a package for a day event (e.g., a football day in some area). On the customer's side, the E-marketplace can provide a facility for customers to establish coalitions (perhaps on the fly) based on current locations (i.e., proximity) and common interests.

Using location technology in mobile commerce (M-commerce) is not a new idea. We have already mentioned some examples, including location-based advertising, in which advertising is targeted based on not only user profiles but also the user's location and time. Other applications are geoinformation and route finding (e.g., Where am I now? Where is the nearest ATM? Where is my friend now?) (e.g., Tewari et al., 2000), and intelligent shopping assistance that provide suggestions, are sensitive to the user's location (e.g., Fano, 1998), and provide services triggered by proximity (Troel, 2001). Further examples of work to support shoppers anytime anywhere, some using intelligent software agents, include Impulse,[9] MyGROCER[10] (Kourouthanasis et al., 2002), location-based reverse auctions (described later), mobile online auctions (Wagner et al., 2002), E-CWE (Maamar et al., 2001), CRUMPET (Poslad et al., 2001), GPS-based location-based services (Jagoe, 2003), Web services in mobile E-commerce (Pilioura et al., 2003), E-parking (Attane and Papi, 2002), wireless advertising (Kolmel and Alexakis, 2002), ad hoc coalition formation, and others as reviewed by Varshney and Vetter (2002). Hence, there is a huge range of services, some of which may be useful for different tasks and at different locations, resulting in the need to organize such services for the user.

In marketplaces, as first noted by Loke and Zaslavsky (2004), ambient services may be classified not only according to geographic boundaries of relevance but also by their suitability to the task within the user's shopping process, when the shopping process model is based on the stages of consumer-buying behavior (He et al., 2003). The context of the user can be extended to include not only current location but also the

---

[9] http://agents.media.mit.edu/projects/impulse/.
[10] http://www.eltrun.aueb.gr/mygrocer/.

stage in which the user is within the Consumer-Buying Behavior (CBB) model. We can use the seven-staged CBB model as additional contextual information to classify ambient services, though some services might span several stages. The seven stages, which typically proceed sequentially, are described in the following text:

Need identification: In this stage, the consumer identifies a need for a product or service. Typically, this need can be stimulated by advertisements, natural means (via conversations with other people), some user-employed suggestion services, and peer recommendations. In the wireless environment, location-based advertising based on proximity to shops will fit into this stage, together with proximity-based tractor beam offers (as mentioned earlier). Otherwise, the need may simply come from the consumer himself or herself (e.g., the consumer feels thirsty or hungry).

Product brokering: In this stage, the consumer determines what to buy to satisfy the identified need. Based on the consumer's criteria, a set of desirable products is created as a result of this stage, with information for evaluating product alternatives. In the wireless environment, various kinds of product comparison services may be invoked, or, if permitted, the Web can be accessed for information required to make an informed decision. Given a user's criteria (including current location), services can be created to provide product recommendations to the user from shops in the neighborhood. Friends in the neighborhood may also be contacted to offer information about purchase suggestions. The consumer may also send queries about products available to shops in the neighborhood, for example, a virtual designer-jeans shopping mall formed by a group of shops in the consumer's surroundings. If such a service is available, the consumer may be able to point to the product with the device, ask questions about its price, make, etc., and look it up in his or her cupboard to see if there are other matching products, as in the case of clothing (e.g., Gershman, 1999). The consumer may also be able to show the product to friends via the device's built-in camera.

Buyer coalition formation: Coalitions among buyers might be formed on the fly to provide greater bargaining ability, but this will require a service for buyers to communicate with each other about their buying intentions.

Merchant brokering: In this stage, the consumer determines who to buy the product or service from. This stage involves merchant or vendor evaluation rather than product evaluation. It is possible to utilize factors such as proximity of consumer to the shop, as well as delivery options — the user might decide to pick the item up himself or herself. Other factors taken into consideration, even in the nonmobile case, are price, reputation of merchant, and, perhaps, long-standing relationships.

Negotiation: In this stage, the consumer discusses with the chosen merchants to reach an agreement on the terms and conditions of the transaction. Such negotiation can occur one to one or involve multiple parties. For example, auctions may be carried out between multiple consumers for a product that a merchant is selling, or reverse auctions may be carried out between a consumer and many selected merchants.

Purchase and delivery: In this stage, the consumer selects among payment options as acceptable to the merchant and makes payment. The delivery of the product is then performed.

Product service and evaluation: In this stage, the consumer provides feedback on the product or service to the merchant or product creators. Such feedback might be immediate or after a period of time.

The consumer can notify a system of the stage he or she is currently at and, in this way, enable the system to suggest services relevant to the specified stage. For example, if consumer Jane is in the need identification phase, she can opt to receive wireless advertising about shops in the neighborhood or, say, in the shopping mall, as well as suggestions from friends in the neighborhood, if any, about what Jane might be interested in. Suppose Jane identifies a need. Then, in the product brokering stage, she might utilize search services to find out where she can find particular items nearby or send a query to friends who might recommend a suitable product or identify the neighborhood shop in which they have seen it. Jane might also use an information service to find information (e.g., price, availability, etc.) of alternative products sold in shops not more than several hundred meters away. Once she has narrowed her options down to one or several possibilities, Jane might then look for a suitable vendor (if not found in the previous stage as answers to product comparison queries) and compare alternatives. Negotiation or bargaining for the product from a chosen vendor might then proceed either via direct human communication or via technology (e.g., Jane makes a call, or her agent automatically negotiates on her behalf while she does something else). If Jane does not know where the vendor is located, she might utilize a navigation service that would lead her not only to the vendor but all the way to the product shelf. Note that Jane is free to change her mind and may decide not to purchase a product or discover that the need is no longer important. Subsequently, Jane organizes to pay for the product or provide a down payment either at the vendor's location or via technology without having to walk to the vendor's premises, and to arrange for pickup.

There will be services that incorporate several stages of the aforementioned sequence. For example, Jane might receive an advertisement concerning a type of Chinese food at a particular restaurant. She might then not have to consider other restaurants that sell the same or alternative food. She decides to take up the offer and walks to the restaurant using the navigation services that came with the advertisement. Another example is the proximity-based reverse auctions, in which the user issues a query such as "Who within a radius of 100 m from me will sell me X for price P?" The system initiates a reverse auction among vendors within the specified area to answer this query; thus, in the course of handling this query, the merchant is determined and the price negotiated. We will later develop this idea of the location-based reverse auctions further.

A consumer might not be shopping for one but for several services or products. For example, a consumer leaves the house for a shopping mall and employs parking services to negotiate and rent parking space on arrival. Thereafter, the consumer reenters the aforementioned stages but for a different product (e.g., tennis shoes). In addition, the consumer might be at different stages concurrently for a number of products or services.

A consumer might become a provider (or when charging is used, a vendor) of a product or service. For example, an individual having some free time can begin to provide information services (e.g., navigational guidance, opinions, or help in product searches) in the context of a virtual community formed by mobile device users in a neighborhood (e.g., Loke, 2002). Moreover, an individual can also sell his or her digital screen estate to vendors who are willing to pay (in terms of monetary units or incentives) to have their advertisements on the user's device (Loke et al., 2004).

As mentioned earlier, we note that further refinement of the consumer's shopping intentions can be made if the consumer also provides the shopping role as described by Stolze and Ströbel (2001). Examples of such roles include "father of a teenage girl," "son for mum's birthday," etc. According to this role, appropriate services can be selected to help the consumer. Such roles help in the need identification stage of the CBB model by conveying hints about the reason for shopping.

The three dimensions described earlier form a service classification space as illustrated visually in Figure 3.3. There are services that extend over different portions of the space when they are relevant in multiple stages, multiple locations, and multiple roles.

3.4.3 Office Building Example

Another example of indoor logical areas and ambient services within an office building, from Loke et al. (2005), is illustrated as a floor plan in Figure 3.4. A typical office building has rooms and various subdivisions.

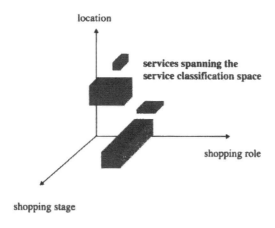

Figure 3.3 The three dimensions for organizing services. The solids represent services within the service classification space.

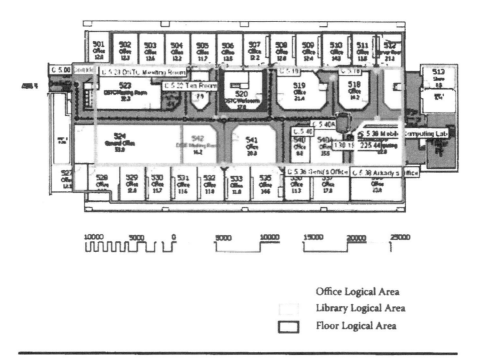

Figure 3.4 Floor of a building in a university campus.

Services can be structured according to various functional subdivisions within a floor of a building. Certain library services, such as navigating to the bookshelf on which a required book can be found[11] and interacting with the library management, can be offered to users with a connected mobile device within the library area. Services offered to users with a mobile device related to office space might include printing documents (from the mobile device to the nearest printer), navigating to a person's office, leaving electronic notes for someone not in the office, accessing various devices in a room via a mobile device (used as a universal remote controller, but while the user is in the room), accessing wireless Internet or Voice-over-IP communication services (e.g., via the WLAN).

3.4.4 A Software Architecture for Ambient Services

To illustrate a possible implementation of ambient services, we outline the architecture and key components of a system, first introduced as the prototype developed in Loke et al. (2005). Figure 3.5 depicts the architecture of a system for proactive discovery and update of ambient services. The system consists of four major components: the Ekahau Positioning Engine (EPE),[12] the Service Calculation Engine (SCE), Mobile Client Application, and Web Services.

EPE 2.0 is the positioning server that keeps track of mobile users in a WLAN. It detects the mobile users who enter a particular geographic location of each user. The user's current location information is then sent to the Service Calculation Engine (SCE) periodically to calculate which service domains the user is currently in and the composition of services available to each user. The system requires that there is a constant connection between the EPE and the SCE.

The Server Module consists of two main components: SCE and Service Database. The SCE plays the most important role in discovering and updating ambient services. Its main task is to calculate the logical areas the user is currently in and a composition of the sets of services available to the user based on the location information received from the EPE. Every time the user moves to a different logical area, a new set of services is calculated and enabled for the user. The SCE also acts as a server to the client devices that enter a particular logical area (i.e., comes within the scope of some services). To send updated sets of services to the user, it listens to the client device connections continuously. When a user enters the area and connects to the server, it establishes a connection with the

---

[11] http://coolcampus.csse.monash.edu.au/MonashLibrary/— a project of the Cool-Campus Initiative.

[12] http://www.ekahau.com.

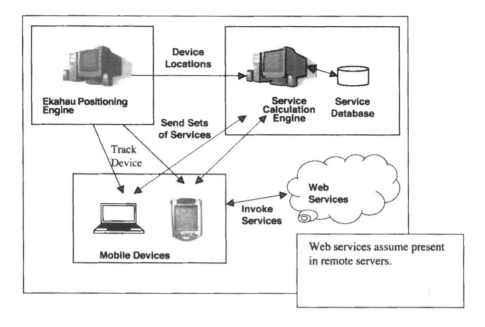

Figure 3.5   Components of a system supporting ambient services.

client device and sends updated sets of services to the client device. As the user moves from one logical area to another, the composition of services available to the user would change, and a new set of services is calculated and sent to the client device. The Service Database is a database repository that stores the details about the logical areas of services. The Service Database is on the same host as the SCE. The SCE retrieves the services from the service database to calculate the locally relevant services at a given geographic area.

The Mobile Client Application is an application installed in the mobile device of the user. This device is tracked by the EPE. The basic function of the client application is to receive the sets of services from the server and display it to the user. When the user invokes the service, the client makes a request to the Web service. The prototype developed is specifically aimed at the pocket PC and developed using the Microsoft.NET Compact Framework (CF). When users enter a particular area, they are required to log in with User name, User type, and Password to check the user type of the mobile user, because different user types have different privileges in the system. After logging in, once the user clicks the Connect button, the mobile device will connect to SCE, which is waiting for client connections on the server machine. It then receives the services that are

Figure 3.6 Services available in area 2 (office logical area) displayed on a user's mobile device.

calculated and transmitted by the SCE. A new set of services is sent to the mobile client application every time the user moves from one logical area to another (as detected by processing data from Ekahau). Thus, the services displayed to the user are updated every time it receives the new sets of services. The services available in area 2 (office area) are displayed for the user in his or her mobile device as shown in Figure 3.6.

For example, when the user moves from area 2 (office logical area) to area 1 (external logical area), a new set of services will be enabled for the user. The services available in area 1 are displayed for the user in his or her mobile device as shown in Figure 3.7.

For evaluation, we use a Compaq iPAQ handheld device and a desktop PC. The Compaq iPAQ was used to implement the mobile client application and host Web service clients. The desktop was used to run the EPE, SCE, and Service Database.

In Figure 3.8, T(logical area 1) is the time a user starts moving from a location. T(logical area 2) is the time the user arrives at the new location. T(message updated) is the time the user sees the new list of services.

The feasibility of the prototype system was measured by calculating the total time taken for the system to discover or update the relevant services for a user once he or she arrives at a new location. The total

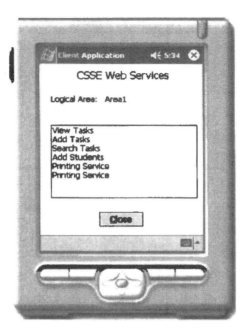

Figure 3.7 Services available in area 1 (external logical area) displayed on a user's mobile device.

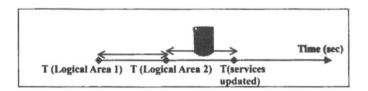

Figure 3.8 Timeline to update sets of services.

time taken T(service updated) is the length of the time between T(message updated) and T(logical area 2). The total time taken T(service updated) is calculated by adding the time taken to track the user's current location, the time taken to access the database, the time taken for calculating services, and the time taken for transmitting data to the mobile device. The formula to calculate T(service updated) is as follows:

$$
\begin{aligned}
\text{T(service updated)} = \ & \text{T(tracking user's location)} \\
& + \text{T(accessing the database)} \\
& + \text{T(calculating services)} \\
& + \text{T(transmission of data between service engine and mobile device)}
\end{aligned}
$$

Table 3.1   Measurement Results

Average Time Measurement	Time (s)
T(tracking user's location)	2
T(accessing the database) + T(calculating services)	5
T(transmitting data to mobile device)	1.2
T(service updated)	8.2

A total of 10 runs for a period of approximately 1 min (60 s) were made to measure the time delay for each of the four processes. Based on the test results, the average time measurements for each process were calculated. T(service updated) is then calculated by adding the time measurements of each process. The measurements are approximate as they depend on the model of network cards, the number of access points, the CPU, and the EPE load. In calculating the overall delay, the EPE tracking delay is also considered, although it does not affect significantly the overall delay. In this testing, the time taken for the invoking of Web services is not calculated as we only focus on the discovery and update of the available list of services at this point. Table 3.1 shows the results. The 8-s delay can be greatly reduced by removing the database access time (i.e., by reading the mapping tables into memory beforehand).

So, the time taken for services to be updated from the previous area to the new area would be (1) the time taken for a user to travel from the previous logical area to the new logical area, i.e.,

$$T(\text{Logical Area 2}) - T(\text{Logical Area 1}),$$

as shown in Figure 3.8, plus (2) T(service updated).

We note that it takes around 2 s to detect a user's change of location, i.e., T(tracking user's location) = 2 s.

This is a characteristic of Ekahau, although such a time lag (either more or less) is inevitable regardless of what positioning technology is used. Other forms of positioning technology (e.g., detecting presence based on a RFID tag worn on a person [Varshney, 2002]) might permit a user's new location to be detected much faster. T(service updated) translates into user wait time. The user wait time can be stated more generally as the time since a change in context (in this case, the service domains the user is located in) or the time the user waits before seeing an updated list of services after he or she has changed context, and is given by the following expression:

$$T(\text{user wait time}) = T(\text{context change detection})$$
$$+ T(\text{compute services for current context})$$
$$+ T(\text{transmission of services updates to mobile device})$$

An implication of this is the effect of the rate at which the user changes context. If the user is moving quickly (e.g., in a bicycle) in and out of service domains, either some of the context changes will not be detected, or, even if detected, the system will lag behind the user. In general, the user should remain in the same context for a period larger than *T(user wait time)* for the system to catch up with the user. This is not a problem if the user needs to stop to invoke and use services.

It is also to be noted that Ekahau has a 1-m accuracy error, which imposes a limitation on the size of logical areas we can have and the extent of movement detectable. Logical areas of too small a granularity are not distinguishable by Ekahau. In other words, visually, the geographic boundaries of ambient services in this case are "thick lines." This means that when the user is standing on such a line, whether the user is still in the service domain or not depends on Ekahau's computation at that time (which in turn depends on the signal strengths at that time). Finer-grained positioning technologies might overcome this barrier.

The Mobile Service Toolkit (MST) (Toye et al., 2005) implements the idea of *site-specific services*, or services that reside in a specific location. The MST client software runs on users' smartphones and supports Bluetooth connection to application servers. Personal information stored on the smartphone can be used to personalize services, and there is user control over the disclosure of such information. An example given involves booking a specific place in the queue of a restaurant and being SMSed when the table is ready. Another example of a service involves interactive advertising in which a user can transmit personal information from the phone to request a brochure about a vacation site.

3.5 FROM AMBIENT SERVICES TO PLACE-BASED E-COMMUNITIES

Short-range networking technologies and ambient services form the basis for place-based electronic communities or PBE communities, for short (called location-based E-communities in Loke, 2002), induced by a location and the (possibly temporary) geographic collocation of people. A PBE community carries the idea of richer member participation and interaction via particular services. Services can be conceptually grouped into those for a given PBE community. We use PBE community here to mean a collection of geographically relevant services and the participants. For example, the PBE community of a hotel is supported by a WLAN for hotel guests and

Figure 3.9 Layered model for PBE communities.

hotel management and services built on top of the WLAN infrastructure. Services might be provided to individual users, without the users necessarily interacting among themselves (e.g., each user can do Web browsing but is unaware of other people connected to the same WLAN).

Such PBE communities constitute virtual spaces, which are counterparts of real-world location spaces, particularly for users on the move. An individual in a hotel has not only the hotel as a physical community but also a corresponding PBE community supported by WLAN. The individual is able to access services or interact with other individuals in the physical (and virtual) community. A PBE community can thus be considered an E-community that is induced by, and is a counterpart of, an individual's surroundings or location. We see the PBE community as, roughly, the highest abstraction of the layered model shown in Figure 3.9.

Note that the individual might still have access to virtual communities on the Web (which are geographically transcending and location-independent) and might only be at a location for a short time (and so have only minimum or rapid interaction with a PBE community).

Rheingold (2002) discusses the phenomena of smart mobs and mobile virtual communities, which can form in an ad hoc fashion with each individual having a mobile device interconnectable to others' devices via short-range device-to-device ad hoc networking or via wide area networks. For example, the use of SMS messages can be employed to rapidly organize meetings at a given place. We see PBE communities as aggregations of services offered via WLAN hot spots.

A supporting software infrastructure can support user interaction within PBE communities. We assume that to access the services of a community, there is a community server.

3.5.1 Interaction between User and Community Server

A multiagent architecture for this purpose is described by Rakotonirainy et al. (2000). We give an overview of this architecture here. Figure 3.10

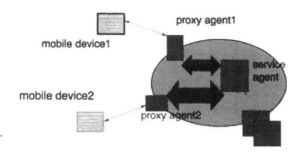

Figure 3.10   Interaction of mobile devices and users within a PBE community supported by a multiagent system.

illustrates two mobile devices that interact with agents acting as proxy for these devices in a PBE community. The proxy agents mediate between mobile device users and service agents. Mediation can mean filtering input streams from devices or providing interfaces to services for mobile device users. The collection of proxy agents and service agents form a multiagent system that supports users in a PBE community. Predefined protocols are used for exchange of messages whenever a device (and user) first joins a PBE community and whenever a device leaves a PBE community. Depending on the underlying networking technology used, the joining protocol can be based on a push technique whereby the community server pushes an "invitation to join" message to the device. Alternatively, a pull technique can be used whereby the user sends a "request to join" message to the community server. The leaving protocol determines what happens when a device (and user) leaves the logical area. With PBE communities, we expect the user to leave the community when leaving the area, albeit this is not necessary (because participation in a virtual community can be retained even if the user is outside the logical area, provided networking technology and software infrastructure supports this). The leaving and joining protocols might be community dependent, varying between communities. Services available can be the ambient services described earlier, as well as services that support community interaction, such as ICQ-like presence technologies or blog spots.

3.5.2   Interaction between User and Multiple Communities: Impact of User Movement

As the user moves across different logical areas, the user also moves across different place-based communities corresponding to those areas. A conceptualization of the user's movements relative to these communities is the *community stack*. An example illustrates this idea.

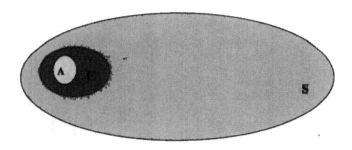

Figure 3.11 Nesting of café, shopping complex, and suburb.

Consider a person P, a member of company X. P enters suburb S (say, in a car), then enters shopping complex C (say, walks in after parking the car), and, finally, walks into café A within C. The nested areas are depicted in Figure 3.11. Then, when P is in A, services available to P can include those in A, C, and S (and even compositions of these services); i.e., P is in the place-based communities for A, C, and S. As P exits A, the services reduce to those in C and S, and as P exits C, the services reduce to those in A. Generalizing from this example, a stack of communities is conceptualized for a user. Each entry into a community (or logical area) pushes the community onto the stack, and each exit from a community (or logical area) pops the community off the stack.

3.6 ENHANCING CONTEXT-AWARE MOBILE SERVICES WITH MOBILE CODE AND POLICY: THE MHS EXAMPLE

Although E-services can be implemented via Web service technology, additional mechanisms are useful for mobile services driven by context awareness, which would need to be delivered to users in an ad hoc manner without the assumption of an extensive *a priori* setup. This section introduces two mechanisms that can be used to enhance the delivery of context-aware mobile services. To illustrate the use of these mechanisms for context-aware mobile services, we describe the Message Hanging Services (MHS) implementation here based on the work of Syukur, Cooney et al. (2004) and Syukur et al. (2004a). One can note the similarity of the MHS framework with the framework for ambient services described earlier; the difference is in that the MHS uses mobile code and policies.

Sensing the user's context (i.e., a user's location), the MHS system can proactively discover and compute a list of services that may be useful to the user at that particular context. This list is sent to the user's mobile device. MHS views a service as a software tool delivered to the user's device for the purpose of suggesting or helping users to complete their tasks. As the user selects from the list a particular service name on the mobile device, the highly compact (with respect to limited device resources) mobile code that provides control for the service is then downloaded. Such a mobile code can encapsulate a user interface and logic to interact with embedded devices or applications in the environment. The mobile code also encapsulates the protocol for interaction with the services on remote servers; because the code is downloaded, details of the protocol need not be on the device beforehand. The combination of context sensitivity and mobile code can provide useful services to the user with minimum or no effort for service setup prior to use.

Apart from applications exposed as Web services, we also consider how the MHS framework can be used to "wrap" traditional applications up as context-aware mobile services. Here, *traditional application* refers to a primitive stand-alone application that does not have or utilize any context-sensing ability. One sample traditional application that will be discussed here is a Windows Media Player application. Adding context sensitivity to a traditional Windows Media Player application enhances the user's experience in using the Windows Media Player.

Apart from adding context sensing, we can also allow the end user to define a policy that specifies what type of music, as well as when and where, he or she wants played or stopped at each particular situation. The policy language here is used to govern and restrict the behaviors of the services according to the user's needs. Another advantage of having a policy is to enable the user to specify tasks to be done automatically in a certain situation (e.g., automatically starts the music at 9 p.m. at room A), which is especially useful if there is regularity in the user's activities. In addition, MHS enables the user to access the Windows Media Player application and control the music from a mobile device.

3.6.1 MHS Architecture

The MHS system provides an infrastructure that mediates the interaction between the client device and the application logic via Web service calls. This high-level architecture is illustrated in Figure 3.12.

One can view the MHS architecture as an extension of the framework for ambient services given in Section 3.4, with additional components to support mobile code delivery and policy processing. Six of the most important components of the system are described here, five as follows and the sixth in the next subsection:

Policies interpreter: This component computes a set of useful services for a user by interpreting the user's policy documents. The interpretation is done on the server side and it takes into account

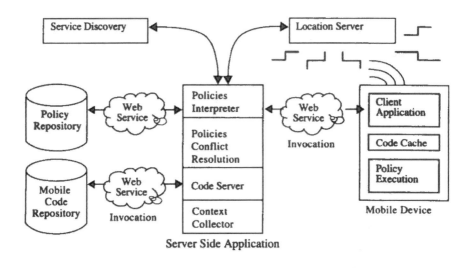

Figure 3.12 High-level architecture of Mobile Hanging Services.

information regarding the user's current contexts, i.e., a user's current
location, day, time, and a user's identity.

Code cache: It refers to the mobile code application stored on the
device for future reuse. Code caching is important if we are using
a code downloading technique.

Code server: This refers to the mechanism that handles a user's request
(from a mobile device), responding to that request by transferring
the relevant mobile code.

Client application: The application resides on the mobile client side
that manages the incoming mobile code and executes the imported
service interface code on the device.

Context collector: Context collector collects all users' contexts information, which are related to the user's current activities. In the prototype
described by Syukur et al. (2004a), the context collection process
is performed by calling the respective Web service. For example, to
get a user's current location, the context collector needs to invoke
a location Web service method.

3.6.1.1 Policy Conflict Resolution

As each entity (e.g., a user) in the system is allowed to create its own
policy and each entity may have a different set of rules, there is a chance
of conflicts occurring. The policy conflict resolution component handles
conflicts between entities in the system, if any. There are three possible
techniques to resolve conflicts. To decide the technique to be used, a

role-based method is employed. The policy conflict resolution component analyzes the type of role that the user has. Role here refers to the level of privilege of an entity in the system (i.e., a general user, power user, or super user). Basically, the purpose of the role is to group and assign different levels of authorities and privileges to each entity.

The grouping here is based on the type of entity. For example, an entity with the type student will have a general user role, a lecturer entity has a power user role, and the head of a school has a super user role. A super user is at the top of the hierarchy in the system, followed by a power user and a general user. An entity with a higher role can do more things compared to an entity with a lower role. For example, a super user can choose either to stop the currently playing music or to play his or her own favorite music. A power user can only stop the music, and a general user is not allowed to stop the currently playing music.

The conflict resolution techniques are discussed as follows:

- Hierarchy override policy: By default if the conflict occurs between users that have different roles, a user with a higher role can override the policy that belongs to the user with a lower role.
- Soft/hard rules override policy: Policy conflict may also occur between users that have the same roles. If such a conflict occurs, the system then needs to detect the type of policy rules that resulted in the conflict, i.e., a soft or hard rule. The policy with a soft rule characteristic is a flexible policy that can be modified depending on the user's current situation. In contrast, hard rules cannot be modified. If both conflict policies have different types of policy rules (i.e., one is a hard rule and the other is a soft rule), then the soft/hard rules override policy will be applied. This conflict resolution technique means that hard rules will always override soft rules.
- Merging policy: The merging policy is used if the conflict occurs among users with the same role and the same type of policy rules (soft rule–soft rule or hard rule–hard rule). Merging here will combine the rules from each entity involved in the policy conflict.

3.6.2 Context-Based Policy Control of Media Player Service

A Windows Media Player is a stand-alone application that will not perform any actions (i.e., start, stop, pause, or resume the music) until there is a request from users to do so. Adding context sensing to any traditional application can improve the user's experience. Using the MHS framework, context-sensitive behaviors can be added to almost any existing traditional application. This is possible as the system permits a remote Web service

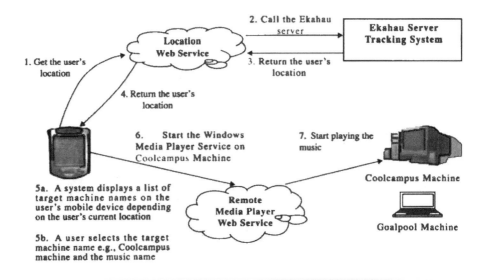

Figure 3.13 Starting the Windows Media Player service from a mobile device.

call from a client device to a server or vice versa and from a server to any computing devices (i.e., a desktop machine). The .NET remoting mechanism is used to access the application's API (Application Programming Interface). Through the API, the MHS system can control the execution of the application.

The ability to download and execute in an ad hoc fashion a mobile code on the mobile device where the code can be used to control the Windows Media Player application (i.e., the user can choose to start, pause, resume, or terminate any music at any target desktop machine[13] that he or she likes from a mobile device and from any location that the user is in) gives the user convenient control over the application.

Figure 3.13 illustrates in more detail how to start the music service manually on the target desktop machine from a user's mobile device. Each step required to start the music on the target machine is discussed as follows:

Steps 1, 2, 3, and 4: Get and return the user's location. Once the user has successfully logged in, the system calls the location Web service to retrieve the user's current location. If found, the information is

---

[13] Although we used a desktop computer in our prototype, one could also imagine a music service server that is not a desktop computer but "faceless" hardware embedded into the environment (e.g., embedded into the wall of a room), and, so, the only means to control the service is the user interface provided via the mobile code on the user's mobile device.

returned to the Windows Media Player client application that resides on the mobile client side.

Step 5a: Display a list of target machine names. As soon as the user's location is returned, the system then displays a list of target desktop machines in that particular location. We store information regarding the mapping between the room and list of machines in an XML database. Knowing a user's current location, the system will be able to search and get a list of available machines for that particular room.

Steps 5b and 6: A user selects the target machine name and starts the Windows Media Player service on the Coolcampus machine. Once the user selects the target machine name (i.e., Coolcampus machine), a list of songs, which are available on that machine, is then displayed. Then, when the user chooses to start, pause, resume, or stop this song, the system then calls a Remote Media Player Web service to execute the music on the Coolcampus machine as specified by the user.

Step 7: Start playing the music. If the user chose to start the music, the Remote Media Player Web service then contacts the remoting client that resides on the Coolcampus machine to start playing the user's music selection.

The steps just mentioned are for starting the Windows Media Player service process manually, i.e., as initiated by a user selecting the music name from a mobile device. Given control on the execution of the application through its API, one can make the Windows Media Player application react more "intelligently and autonomously" (without the user's intervention) by integrating some context information about the user and extra rules or policies that specify when and where to automatically start and stop the music service in response to context changes. For example, a user's favorite music is automatically started at the user's desktop machine as soon as the user enters room A, and once the system detects that the user has walked out of the room, the system will then pause or stop the music.

After presenting the utility of having policy documents in the context-aware ubiquitous environment, we note design criteria for a policy language. The first consideration is whether the policy document is easy to be used and understood by users. *Easy to use* here means that the policy language is simple (not overly complicated) and all notations or element names are largely self-describing. The second consideration is that the policy language should be extensible and capable of being used in any situation (simple or complex situation). In our definition, a simple situation consists of only a single entity (i.e., one user) at the particular space and time. A complex situation here involves two or more entities with different sets of rules or policies. In complex situations, a conflict is more likely to occur as different users may have different rules to govern the service execution for the same context.

Taking into account design criteria mentioned earlier, MHS currently provides a simple initial design of a policy language with eXtensible Markup Language (XML). The policy language is structured according to the user, i.e., a user name, followed by day, location, and start-and-end time when the policy needs to be executed. For example, by looking at the following sample policy document, we can see that the user named "Bella" wants to start the "Secret Garden" song automatically at her desktop machine from 2 p.m. to 3 p.m. on Sunday.

```xml
<?xml version="1.0" encoding="UTF-8"?>
<mediaPlayerPolicy>
  <user name="Bella">
    <policyDetails>
      <activity day="Sunday">
        <location name="MobCompLab">
          <activityDetails startTime="2:00PM" end
          Time="2:30PM">
            <action type="start">
              <service name="Media Player Service">
                <songName>SecretGarden.wav</songName>
                <machineIPAddress>130.1.194.224
                  </machineIPAddress>
              </service>
            </action>
          </activityDetails>
        </location>
      </activity>
    </policyDetails>
  </user>
</mediaPlayerPolicy>
```

To process policies, the system uses multiple threads. The threading process will run once the mobile client application is started on the device. This running thread will keep monitoring the current time and location of the user. Once the time and location are detected, the system then automatically transfers and executes the specified service code on the mobile device. After this, the steps are the same as steps 6 and 7 mentioned earlier.

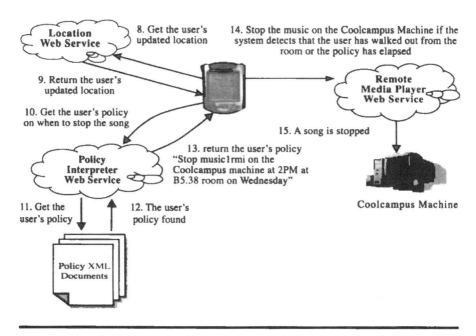

Figure 3.14 Stop the Windows Media Player service on the target machine.

In general, there are two possible reasons the system needs to stop the song automatically on the target machine: (1) the system detects that the user is no longer in the room; i.e., the user has walked out of the room, and (2) the time to stop that particular music in the user's policy file has been detected. A user can also manually stop the music by clicking on the stop button on the Media Player service interface from a mobile device.

By specifying such behaviors in a policy file, the user is freed from performing regular tasks, i.e., starting and stopping the music service, and relies on the system to do it automatically.

After describing the steps required to start the music on the target machine, we discuss the steps to terminate the music (see Figure 3.14):

Steps 8 and 9: Get the user's updated location and return the updated location. The system first checks whether the user is still in the same room. If the system detects that the user has walked out of the room, the system then continues with Step 14. Otherwise, steps 10, 11, 12, and 13 are performed.

Steps 10, 11, 12, and 13: Get the user's policy and return the policy to the mobile client application: If the system detects that the user is still in the same room, the system then looks for the user's policy. If the time to stop the music has elapsed, the system then continues with Step 14. Otherwise, the system continues to monitor the user's current location (refer to steps 8 and 9).

Step 14: Stop the music on the target machine. If there is a request to stop the song, the remote client that resides on the target machine then needs to invoke a stop method to terminate the song playing on this machine.

Step 15: A song is stopped. On invocation of the stop method, the Windows Media Player service stops playing the music.

Steps 1 to 15 described in the preceding text can be the same for adding context awareness to other applications such as Virtual Network Computing application, Games, etc. The difference is that instead of the Remote Media Player Web service starting the Media Player process on the target machine, another type of application process will be started. The user may want to specify different execution and termination policies for different applications in the user policy XML document.

A performance evaluation of the MHS prototype that involves measuring the time it takes to get a user's updated location to the time it requires to execute the service on the mobile device is given by Syukur et al. (2004b). Different heuristic techniques can be used to improve the MHS performance, i.e., by reducing service execution time and context change delay when the user moves from one place to another. The results show that the user wait time is not prohibitive and highlight the importance of runtime efficiency for context-aware systems.

3.6.3 Partial Control between User and System

As the MHS system supports both manual and automatic execution of services, it is important to clearly separate the control between users and systems; i.e., when should control be given to the user and when should the system be in control? Giving control back to the user becomes extremely important, especially when the user performs some odd activities during the day, which is different from the tasks that he or she has specified in the policy document. For example, a user has a group meeting at room A (the user's office). He specifies in the policy document to start the music at his office at 3 p.m. (i.e., after the meeting). But, what happens if the meeting is not finished at 3 p.m.? Given a policy rule, as soon as the current time is 3 p.m., the system will automatically start the music in room A. The system might do this automatically by interpreting a user's policy document, unable to tell accurately whether the meeting is over or not.

If such a situation happens, most likely the user will want to manually stop playing the music from his or her mobile device. This is done by selecting the service name, i.e., Media Player, on the mobile device, and a mobile code with respect to this Media Player service will then be downloaded to the user's mobile device. Once the service interface is displayed, the user then clicks on the stop button to manually terminate the music. Once the system detects that the user is manually performing the task and this task is different from the activity specified in the document, full control is returned to the user. The system will not perform any further policy interpretation (and music execution) until the system detects that the user has closed the Media Player service form. Once the form is closed, full control is returned to the system. The system then continues to interpret the user's policy document and automatically starts, pauses, resumes, or stops the music. In summary, MHS's current control scheme is as follows: the user takes control of the service by requesting and using the mobile code application (containing the user interface) for the service, but control is returned to the system when the user closes this service application.

3.6.4 MHS Summary

The MHS prototype described earlier can be extended, for example, by including more complex policy, which take into account conflicts between entities in the system, and more types of contexts such as a history log file and physical sensors. As presented here, MHS illustrates how mobile services can benefit from deployment of mobile code and how the concept of policy is useful for specifying context-aware behaviors.

3.7 ENHANCING CONTEXT-AWARE MOBILE SERVICES WITH MULTIAGENT TECHNOLOGY: THE EXAMPLE OF PROXIMITY-BASED REVERSE AUCTIONS

We have already seen an application of multiagent technology to support place-based communities. This section further explores the idea of using intelligent software agents to enhance physical marketplaces, as they are employed in the context of assisting shoppers (i.e., a shopper in a shopping mall or a popular shopping street such as Orchard Road in Singapore, which is lined with shopping complexes, or Bourke Street Mall in Melbourne, Australia). More specifically, we consider using agents to answer a specific kind of user query with a concept called *proximity-based reverse auctions*, which represents an E-marketplace superimposed on a physical marketplace. This section is based on the work of Loke (2003).

3.7.1 Proximity-Based Reverse Auctions

In reverse auctions, instead of buyers bidding for the goods of a seller, sellers bid for a buyer, who announces a price he or she is willing to pay for some item. The reverse auction helps to answer the buyer's query of the kind "Who can sell me X at price P (or less)?" Proximity-based reverse auctions are reverse auctions with constraints due to the user's location and time limitations. A typical query that such auctions will attempt to answer is, "Who can sell me X at price P (or less)? And without me walking further than 100 m from here, and I would like to know this within 3 min." We call such queries *PRA queries*. Such queries can be issued from a user's mobile device to a stationary Bluetooth access point or a WLAN access point (on which runs server software to which vendors are connected) while the user is, say, in a shopping mall or on a street, or via a GPS-enabled device.

We can also view proximity-based reverse auctions as taking a step further than the ordinary location-based queries of "where can I get X?" to "where (and how) can I get X for price P (and tell me the answer within the next T minutes, and give me an answer where I do not need to walk too far)," where P can be determined dynamically by a reverse auction.

Implicit in the query are three constraints: (1) the location constraint — X should be bought from a shop not too far from where the user currently is, (2) the time constraint — the query should return a reply to the user within several seconds or, at most, several minutes because the user (on the move) would not be at the same location for a long time, and (3) the price constraint — the query should return an answer within the stated price, or near the stated price. If no such price is found, then return the lowest price found.

There might be a trade-off among constraints that could be prespecified by the user. For example, a nearer shop might sell X for a slightly higher price than a shop further away, and the user might choose the nearer shop if the price is not too much higher. The ideal answer to the user's query is one that is nearest and cheapest. But if the ideal is not possible, the answer might comprise a set of alternatives from which the user can choose. As soon as a query is issued, multiple vendors are triggered to be involved in a reverse auction to bid for the user's sale. The duration of the auction is a factor here; if the user cannot wait, and the lifetime of the auction is too short, the resulting price might not be optimal.

The concept of reverse auctions is not new, and corresponds to a shift of trend from vendors asking "would you like this?" to buyers asking "who can get me this?" Proximity-based reverse auctions is also perhaps not a radically new concept; an example: several parking facilities or family restaurants within an area can bid for a car's business. There has also been recent work in developing middleware to bring online auctions from the desktop to mobile devices, and the popular auction site www.ebay.com has enabled access to their auction site via PDAs and SMS messaging.

An issue with such auctions in which participants or potential participants are mobile and only have resource-limited devices is how to monitor the status of auctions. For example, one may be interested in what auctions are going on, what is happening in a particular auction, or the results of recent auctions for a particular type of product but does not have the luxury of browsing an informative data-intensive auction Web site because of device limitations and being on the move. One way to get such information is to use a publish/subscribe event notification system in which a user subscribes to auction events of interest, and event notifications generated by the system are forwarded to the user as small chunks of information over time. Event-based (or push-based) observation might fit better into the mobile computing context than browsing-based (or pull-based) observation, for both users and agents, because of potentially better bandwidth utilization than in continuous polling and the asymmetry in the mobile environment (a lot more mobile clients than servers, greater need to conserve power on the client side, less computational and memory resources on the client side, and higher server-to-client bandwidth in some types of wireless networks). But more on such an approach to reverse auctions later.

3.7.2 A System for Proximity-Based Reverse Auctions

3.7.2.1 Architectural Overview

A multiagent system is presented for proximity-based reverse auctions to automate most of the process. Such a system will provide invaluable aid to both vendors (shops) who might service hundreds or thousands of PRA queries at a time (and so be involved in many reverse auctions at the same time), and buyers who might initiate several such PRA queries at the same time and who are on the move. There are many possible architectures for such a system, but we contend that it should at least comprise four types of agents: user agent, broker agent, vendor agent, and observer agent. Figure 3.15 illustrates such an architecture. The user agent runs on the mobile device, whereas the other agents run on the server side on stationary hosts. An outline of the agents' functions is as follows:

- User agent: The user agent runs on the user's mobile device and interacts with the user, mediating between the user and the server-side agents. The agent accepts PRA queries from the user (i.e., the potential buyer), expands it with the user's preferences stored locally, and then forwards the query to the broker agent over a wireless connection. The user agent also accepts event subscriptions from users and forwards them to observer agents. Event

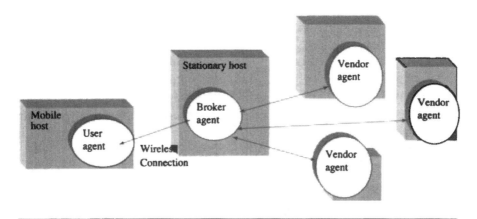

Figure 3.15 **The multiagent system for a proximity-based reverse auction spawned by a mobile user.**

notifications are also received by the user agent and displayed to the user.

■ Broker agent: The broker agent runs on a stationary server receiving PRA queries from the user agent. On receiving a PRA query from a user agent, it invites vendors (represented by their agents) to participate in a reverse auction to answer the PRA query (i.e., to find a vendor with the lowest price satisfying the time and proximity constraints). The broker agent then coordinates the reverse auction playing the role of the auctioneer.

■ Vendor agents: A vendor agent represents a vendor in a reverse auction, including bidding and settlement. The strategy used by the agent would be determined by the vendor. It is the vendor agents (representing different vendors) that bid against each other for the purchase.

■ Observer agents: Observer agents act on behalf of users. On receiving subscriptions from user agents about particular events, the observer agent subscribes to the requested events (with an event notification system) on behalf of the user.

Note that additional types of agents can be included in a more sophisticated architecture with greater functionality or to improve efficiency, or many instances of such agent types might be employed. For example, a vendor might use hundreds or thousands of agents — one to participate in each auction — or one agent to participate in several auctions at the same time. Also, there could be separate agents for the broker and auctioneer roles and a hierarchy of brokers over hierarchically structured geographic domains; for example, there can be a broker for a shopping mall, a broker for each shopping complex in the mall, and a broker for each department of a shopping complex. However, we will not focus on such implementation details here but on the least of what such a system would need (i.e., the minimum components to achieve the required functionality).

As noted in Loke (2003), a prototype proof-of-concept system with minimum functionality using the JADE-LEAP toolkit[14] has been implemented. The system consists of a user agent running on a Palm Vx emulator, a broker agent, and vendor agents, all running on a WinME laptop. Note that the vendor agents and broker agents can run on different machines, which is what one might expect in a real-world deployment.

In a typical run, the user poses a query with time, distance, and price constraints to the user agent, which then passes the query on to the broker agent. The broker agent then selects vendors within the given distance constraint before initiating a reverse auction with the selected vendors' agents. Adaptation of the English auction FIPA protocol is used to perform a reverse auction bounded by the user's time constraint. The auction will end when time is up or, after a wait time, when there are no other competing bids. The user is then informed of the vendor with the best prices and locations attained.

3.7.2.2 From the User's Viewpoint

Figure 3.16(a) shows the JADE-LEAP user agent on start-up. The user interface contains a list of products that the user can select from and indicates the price required. Figure 3.16(b) shows a feature of the user agent that enables the user to configure his or her profile. The profile information includes the time the user is willing to wait for the results of the query (this imposes a time bound on the whole auction), the maximum distance between the user's current location and the vendor, and vendors that the user does not want to buy from. The profile information is included with the user query sent to the broker agent. The broker agent, on receiving the query, initiates bidding among vendors and, after bidding has ended, forwards the results to the user agent. Figure 3.16(c) shows the results of an auction showing the product the user wanted to purchase, the lowest price (satisfying the user constraints), the vendor selling the product with that price, and the distance of the vendor from the user. Figure 3.16(d) shows the receipt the user receives after the user confirms the purchase.

---

[14] http://jade.tilab.com/.

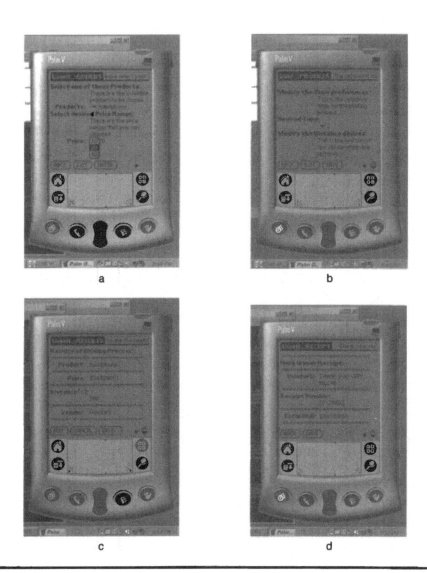

Figure 3.16   (a) The user agent starts up and provides an interface for the user to enter a query. (b) The interface for updating user profile information. (c) The results of an auction to buy a hand phone. (d) The receipt after the user clicks on the "buy" button in Figure 3.16(c).

3.7.2.3 Interaction Protocols

The interactions among the user agent, broker agent, and vendor agents are captured in four interaction protocols, one for each stage described as follows, with the waiting time specified by the user allocated to the second and third stages:

1. User and broker (user initiates purchase): The user is in control and hence no part of the user-specified waiting time should be allocated to this protocol. This protocol is based on the FIPA Brokering Interaction Protocol Specification.[15]

2. Broker and vendor (system initialization/vendor selection): Because the user is not in control, a certain percentage of the user-specified waiting time should be allocated to this protocol. This protocol was designed to handle the initialization of auctions. This is used after the user has made a product request and involves the broker and vendor agents establishing connections. The broker agent uses the list of undesired vendors to filter out vendors (and also filters out vendors who are too far away according to the user's distance constraint) and then attempts to establish contact with the vendors (in its database or obtained by querying a registry) it considers viable and "worthy" of its time and who might want to participate in the reverse auction.

3. Broker and vendor (bidding protocol): Because the user is not in control, a certain percentage of the user-specified waiting time should be allocated to this protocol. The waiting time is used as a time bound for the auction. Note that the time imposes constraints on the strategies vendor agents use in the bidding. The time affects agents in the sense that if too little time is given, bad choices may be made by the agents, and, if there is too much time, the auction would proceed at a less-than-normal pace. The protocol implements a reverse English auction based on the FIPA English Auction Interaction Protocol Specification.[16]

4. Broker and user (results protocol): The user is in control and hence no part of the user-specified waiting time should be allocated to this protocol. This protocol was designed to handle the finalization of auctions. This is used after the broker has successfully found (via a reverse auction) a vendor who is willing to sell the product to the user. The system waits for the user to confirm the purchase of the product, and, if the payment infrastructure is in place, the

---

[15] http://www.fipa.org/specs/fipa00033.

[16] http://www.fipa.org/specs/fipa00031.

user can complete the transaction. The item can then be made available for collection when the user drops in at the vendor.

The auction proceeds in rounds, in each of which the broker calls for and receives bids, waiting up to a certain time limit for the registered vendors to respond. At the end of this time period, the broker evaluates the bids, responds to the bidders (informing them of the highest current price), and initiates a new round with another call for bids. Note that because there is a time limit, there is a maximum number of rounds. Also, if in a round there are no further bids, the auction terminates. The user's waiting time allocated among the activities in stages 2 and 3 is distributed among the various stages. This includes the maximum amount of time the broker waits for vendors contacted to respond, the maximum amount of time the broker waits for responses to bids (call for proposals) in each round of the auction, the time the broker takes to evaluate the bids at each round of the auction, and the maximum amount of time the broker waits for vendors to reply to the user's purchase confirmation.

3.7.2.4 *Observing Auctions via Auction Events*

The system described earlier augmented by a facility for observing auctions is called the *observer* system. This system allows potential buyers and sellers to view the current items for auction and the various associated attributes. This enables one to "shop around" and see what items are up for auction, but one need not necessarily get involved in the auction. The observer system is based on a noncommercial version of Elvin,[17] an event-based publish/subscribe notification service. Notifications containing information on a certain event are sent to a central Elvin server for distribution to those who have subscribed to the event itself. There are primarily two groups of observers: those who would be buyers and those who would be sellers (i.e., vendors or retailers). Information the buyers might be interested in includes the current best price of an item, items that are for auction in various categories, details of items (i.e., model, features, etc.), the sale of an item (i.e., status of an auction, current price so far in an auction, etc.), various statistics about sales, and similar items but different model and make (which might be cheaper or better). Information the sellers might be interested in includes details of buyers who want items that they are selling, other competitors' actions, various statistics of the marketplace (e.g., quantities of particular items other retailers have auctioned off), and number of buyers in a particular category.

---

[17] http://www.mantara.com/products/elvin-router/.

An observer can subscribe to the system to be informed of the values of the following attributes about an item:

Static Attributes		Dynamic	
Item ID:	string	**Current price:**	monetary
Description:	string	**Time left:**	seconds
Start:	timestamp	**No. bids:**	long integer
End:	timestamp	**Lowest bidder ID:**	string
Start price:	monetary	**Bid history:**	array(bids)
Quantity:	integer	**Auction status:**	string
Detail description:	string		
Image of item:	binary file		
Seller ID:	string		
Category:	string		

Auction status refers to a stage in the auction process. Figure 3.17 shows the various stages in an auction.

For example, potential buyers who are "just looking" can subscribe to the system providing the category of items of interest or partial details of an item of interest (keywords for various attributes) and obtain full details of matching items and current auctions for the items of interest, or ask for statistics about auctions in the area and obtain the number of retailers, the number of items on auction, the number of auctions, the number of categories, the most expensive item so far, and the cheapest item so far. Potential sellers can look for buyers by subscribing to the service to be notified about those who expressed interest in the items they are selling. We expect the broker agents to generate and forward notifications to the Elvin server as auctions are created and executed. Also, observer agents represent users (potential buyers or sellers), storing and forwarding subscriptions and notifications to and from the Elvin server.

Fonseca et al. (2001) describes a system implemented using the Zeus multiagent toolkit for the shopping mall of the future, in which the mobile user engages as a bidder in an English auction involving stores in the mall. This proximity-based reverse auction system differs in the way it answers a particular kind of user query using reverse auctions; vendors are selected dynamically based on the query and user profile information.

Issues of trust and security, as well as how commitments can be represented, have not been dealt with in detail; for example, an electronic receipt for a purchase would need to be validated. These are familiar issues with M-commerce. Dealing with failures and sudden disconnections (e.g., the buyer simply changes his or her mind midway through an

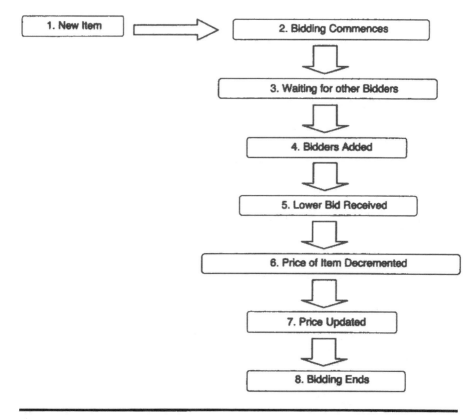

Figure 3.17 The stages an item goes through in the auction system.

auction, or the wireless connection fails) will be areas for future work. The prototype currently uses a fixed allocation of time for vendor registration and the auction process. Dynamic allocation of time might be more useful; for example, if enough vendors responded in a short time, the system need not wait too long before starting the auction.

3.8 SUMMARY AND FURTHER DEVELOPMENTS

We have explored the notion of context-aware mobile services from a number of different perspectives, introducing the concepts of ambient services, PBE communities, and location-based reverse auctions, and illustrating possible implementation schemes with examples of architectures. It is noteworthy that such localized services would tend to complement mobile Internetwide services rather than replace the global Internet services. Service-oriented computing remains an active area of research, and the foreseeable future will see further developments in the area, enabling (and perhaps going beyond) the possibilities mentioned here. Chapter 5 returns to this notion of mobile services but from the perspective of software agents on mobile devices.

ACKNOWLEDGMENT

This chapter contains (1) portions reprinted, with permission, from Loke, S.W. and Zaslavsky, A., Integrated ambient services as enhancement to physical marketplaces, *Proceedings of the HICSS-37 Minitrack on Mobile Distributed Information Systems*, January 2004, ©2004 IEEE, (2) portions from Loke, S.W., Krishnaswamy, S., and Naing, T.T., Service domains for ambient services: concept and experimentation, *Mobile Networks and Applications (MONET)* — special issue on mobile services 10, 2005, Springer Science+Business Media, Inc., pp. 395–404, with kind permission of Springer Science and Business Media, (3) portions reprinted, with permission, from Syukur, E., Cooney, D., Loke, S.W., and Stanski, P., Hanging services: an investigation of context-sensitivity and mobile code for localized services, *Proceedings of the IEEE International Conference on Mobile Data Management*, Berkeley, CA, 2004, pp. 62–73, ©2004 IEEE, and (4) portions obtained from Loke, S.W., An exploration of agent assistance for physical marketplaces: proximity-based reverse auctions, *Proceedings of the International Conference on Intelligent Agents, Web Technologies and Internet Commerce — IAWTIC2003*, Mohammadian, M., Ed., February, 2003, Vienna, Austria, pp. 124–135.

REFERENCES

Attane, M. and Papi, J., E-parking: user-friendly ecommerce to optimize parking space, *Proceedings of the 1st International Conference on Mobile Business*, Greece, July 2002.

Fano, A. Shopper's eye: using location-based filtering for a shopping agent in the physical world, in *Proceedings of the International Conference on Autonomous Agents*, ACM Press, 1998, pp. 416–421.

Fonseca, S., Griss, M., and Letsinger, R., An Agent-Mediated E-Commerce Environment for the Mobile Shopper. Hewlett-Packard Laboratories, Technical Report HPL-2001-157, June 2001.

Gershman, A., McCarthy, J., and Fano, A., Situated computing: bridging the gap between intention and action, *Proceedings of the 3rd International Symposium on Wearable Computers*, San Francisco, CA, 1999.

He, M., Jennings, N.R., and Leung, H.-F., On agent-mediated electronic commerce, *IEEE Transactions on Knowledge and Data Engineering* 15(4), 985–1003, 2003.

Jagoe, A., *Mobile Location Services: The Definitive Guide*, Pearson Education, 2002.

Karp, A., E-speak explained, *Communications of the ACM* 46(7), 112–118, ACM Press, U.S.A., July 2003.

Kolmel, B. and Alexakis, S., Location based advertising, *Proceedings of the 1st International Conference on Mobile Business*, Greece, July 2002.

Kourouthanasis, P., Spinellis, D., Roussos, G., and Giaglis, G., Intelligent cokes and diapers: Mygrocer ubiquitous computing environment, *Proceedings of the 1st International Mobile Business Conference*, July 2002, pp. 150–172.

Leeper, D.G., A long-term view of short-range wireless, *IEEE Computer* 39–44, June 2001.

Loke, S.W., Modelling service-providing location-based E-communities and the impact of user mobility, *Proceedings of the 4th International Conference on Distributed Communities on the Web (DCW 2002)*, Plaice, J., Kropf, P.G., Schulthess, P., and Slonim, J., Eds., April 2002, Sydney, Australia, Springer-Verlag, Lecture Notes in Computer Science 2468, pp. 266–277.

Loke, S.W., An exploration of agent assistance for physical marketplaces: proximity-based reverse auctions, *Proceedings of the International Conference on Intelligent Agents, Web Technologies and Internet Commerce — IAWTIC2003*, Mohammadian, M., Ed., February, 2003, Vienna, Austria, pp. 124–135.

Loke, S.W. and Zaslavsky, A., Integrated ambient services as enhancement to physical marketplaces, *Proceedings of the HICSS-37 Minitrack on Mobile Distributed Information Systems*, January 2004.

Loke, S.W., Zaslavsky, A., and Jain, B., Wireless marketing of ephemeral personal goods: the case of auctioning screen estate for wireless advertisements, *Proceedings of the 3rd International Workshop on Wireless Information Systems*, WIS 2004, in conjunction with ICEIS 2004, Porto, Portugal, 2004, INSTICC Press, pp. 127–133.

Loke, S.W., Krishnaswamy, S., and Naing, T.T., Service domains for ambient services: concept and experimentation, *Mobile Networks and Applications (MONET)* — special issue on mobile services 10, 2005, Springer Science+Business Media, Inc., pp. 395–404.

Maamar, Z., Yahyaoui, H., Mansoor, W., and Vd Heuvel, W.-J., Software agents and wireless E-commerce, ACM *SIGecom Exchanges* 2(3), 10–17, 2001.

Newell, F. and Newell, K., *Wireless Rules: New Marketing Strategies for Customer Relationship Management Anytime, Anywhere*, McGraw-Hill, New York, 2001.

Pilioura, A., Tsalgatidou, S., and Hadjiefthymiades, S., Scenarios of using Web services in M-commerce, *ACM SIGecom Exchanges* 3(4), 28–36, 2003.

Poslad, S., Laamanen, H., Malaka, R., Nick, A., Buckle, P., and Zipf, A., CRUMPET: Creation of userfriendly mobile services personalized for tourism, *Proceedings of 2nd International Conference on 3G Mobile Communication Technologies*, 2001, pp. 26–29.

Rakotonirainy, A., and Loke, S.W., and Zaslavsky, A., Towards multi-agent support for open mobile virtual communities, *Proceedings of the International Conference on Artificial Intelligence (IC-AI 2000)*, Vol. I, Arabnia, H.R., Ed., Las Vegas, NV, 2000, CSREA Press, pp. 127–133.

Rheingold, H. *Smart Mobs: the Next Social Revolution*, Perseus Books Group, U.S.A., 2002.

Singh, M. and Huhns, M.N., *Service-Oriented Computing: Semantics, Processes, Agents*, John Wiley & Sons, U.S.A., 2005.

Singhal, S., Bridgman, T., Suyranarayana, L., Manuey, D., Chan, J., Bevis, D., Hild, S., and Alvinen, J., *The Wireless Application Protocol: Writing Applications for the Mobile Internet*, Addison-Wesley, Longman, 2001.

Stolze, M. and Ströbel, M., The shopping gate — enabling role- and preference-specific e-commerce shopping experiences, in *Web Intelligence: Research and Development*, Zhong, N. et al., Eds., *Lecture Notes in Artificial Intelligence*, Vol. 2198, Springer-Verlag, Germany, 2001, pp. 549–561.

Syukur, E., Cooney, D., Loke, S.W., and Stanski, P., Hanging services: an investigation of context-sensitivity and mobile code for localized services, *Proceedings of the IEEE International Conference on Mobile Data Management*, Berkeley, CA, 2004, pp. 62–73.

Syukur, E., Loke, S.W., and Stanski, P., A policy based framework for context aware ubiquitous services, *Proceedings of the Embedded Ubiquitous Computing Conference*, Aizu-Wakamatsu, Japan, August 26–28, 2004a, Springer-Verlag, *Lecture Notes in Computer Science*, LNCS 3207, pp. 346–355.

Syukur, E., Loke, S.W., and Stanski, P., Performance issues in an infrastructure for mobile hanging services, *Proceedings of the First International Conference on Mobile Computing and Ubiquitous Networking (ICMU)*, NTT DoCoMo R&D Center, Yokosuka, Japan, 2004b, pp. 32–37.

Tewari, G., Youll, J., and Maes, P., Personalized location-based brokering using an agent-based intermediary architecture, in *Proceedings of the International Conference on E-Commerce*, Seoul, Korea, 2000.

Toye, E., Sharp, R., Madhavapeddy, A., and Scott, D., Using smart phones to access site-specific services, *IEEE Pervasive Computing* 4(2), 60–66, 2005, IEEE Computer Society Press.

Troel, A., Banatre, M., Couderc, P., and Weis, F., Predictive scheme for proximate interactions, *Proceedings of the International Workshop on Smart Appliances and Wearable Computing (IWSAWC'01)*, April 2001, pp. 235–239.

Varshney, U. and Vetter, R., Mobile Commerce: Framework, Applications and Networking Support, *Mobile Networks and Applications* 7, 2002, pp. 185–198.

Wagner, M., Balke, W.-T., and Kießling, W., An XML-based multimedia middleware for mobile online auctions, in Filipe, J., Sharp, B., and Miranda, P., Eds., *Enterprise Information Systems III*, Kluwer Academic Publishers, The Netherlands, 2002, pp. 259–269.

4

CONTEXT-AWARE ARTIFACTS

In this chapter, we review examples of context-aware artifacts, including everyday objects, appliances, and handheld devices, and how they can be made context aware. Other labels for context-aware artifacts include the terms *sentient* and *smart,* describing objects. We discuss the concepts of self-supported context awareness and infrastructure-supported context awareness, and how the functionalities of sensing, reasoning, and initiating actions might be distributed between artifact and infrastructure. We also explore in detail an example of a software system for context-aware mobile phones.

4.1 AWARE OBJECTS

An exciting development in context-aware computing is aware objects, aware everyday appliances, and aware devices. The excitement of seeing a robot perform human-level tasks such as walking or playing table tennis can perhaps not be too far behind that of seeing a soft toy effectively come alive, responding and reacting to its user's handling context, or hearing warm greetings from your television or favorite teddy as you come in through the door. Consider walking through shelves of appliances in a departmental store and hearing some of these appliances try to introduce and market themselves as you come near or touch them; imagine cell phones automatically behaving in the right way under the right circumstances, whether in a meeting or a noisy marketplace, or, say, if the call is urgent. In fact, Gatenby (2005) describes a system that detects a user passing near a shelf of books and, with the help of a user profile, sends SMS messages about possibly interesting books on the shelf to the user's phone. Other books on the shelf nearby related to the one being picked up by the user can also light up (the books have small lights attached to them); the books effectively recommend themselves to the user. Context-aware behavior in devices, appliances, and everyday objects are an emerging new experience.

A context-aware artifact is able to perceive the situation of a user and reacts sensibly to it. Two approaches to context-aware artifacts have been identified by Loke (2006):

- *Self-supported context awareness,* in which an artifact is enhanced with hardware (and perhaps software) having the ability to perceive context and utilizes context in its behaviors.
- *Infrastructure-supported context awareness,* in which a device or artifact acquires context-aware capabilities by utilizing hardware and software infrastructure external to the artifact. The infrastructure might be associated with the environment of the artifact and shared by other artifacts and applications.

Self-supported context-aware everyday objects and digital devices involve sophisticated versions of the original artifacts, attached with hardware and software to enable the artifacts to perceive the user or the environment, and include the following:

- A wheelchair with hardware to aid interaction between the user and the environment (Salvador et al., 2005).
- Dietary-aware dining table augmented with sensors to monitor movement of food on the table (Chang et al., 2006).
- Smart couch,[1] which, equipped with weight sensors, determines the weight of the person, identifies him or her, figures how much he or she is moving, and perceives the approximate position of the person on the couch.
- Chairs with sensors to detect a person, sitting or not, and his or her orientation (Nakajima et al., 2005).
- Chameleon tables (Selker et al., 2002) embedded with sensors so that they know where they are being used, how they are being used, when they are being used, and who is using them.
- A tablecloth (from the Equator project,[2] which works on responsive electronic furniture) that can signal how long things have been left on it.
- The smarttable, which has a grid of tabletop sensors to locate and identify objects (e.g., children's blocks in the intended kindergarten application) on its surface (Steurer and Srivastava, 2003).

---

[1] http://www.dsg.cs.tcd.ie/index.php?category_id=350.
[2] http://www.equator.ac.uk.

- The sensetable[3] (Patten et al., 2001), which can sense and track tagged objects (with physical dials to change their states) on its surface.
- Shelf with pressure sensors to detect out-of-stock retail levels (Metzger, 2005).
- Smart furniture, each article with possibly different sensors to detect user activities and hardware to provide access to Internet services for users (Ito et al., 2003).
- Toothbrush and mirror combination with sensors (accelerometer) to detect usage (e.g., proximity of toothbrush with mirror, position of toothbrush with respect to mirror, and state of the toothbrush) (Fujinami et al., 2005).[4]
- The mediacup using temperature and motion sensors to detect the cup's situation such as cup is stationary, drinking out of the cup, cup is played with, cup is carried around, cup is filled up, drink has cooled off, and the current temperature (Beigl et al., 2001).[5]
- The chameleon mug[6] under development at MIT, which seeks to change color, display safety messages, and spring a handle if the fluid in it is hot.
- A plastic pill bottle equipped with RFID tags read by a reader connected to a computer that can be used to sense the pill bottle — alerting the user, if the bottle has not been lifted off its stand, that possibly the person has not taken the medication (Agarawala et al., 2004); a similar idea is the medication uptake sensors and alerts that, depending on where the user is located, are used in the context-aware medication-reminding system (Mihailidis et al., 2003).
- The intelligent spoon at MIT, with embedded sensors,[7] which "seeks to provide information, in an integrated manner, about any food the spoon is in contact with, and to offer suggestions to improve the food. The spoon is equipped with sensors that measure temperature, acidity, salinity, and viscosity, and is connected to a computer via a cable."
- Smart sink, which can interpret users' tasks (as perceived via a camera) at the sink to provide hands-free control of temperature and water flow, and up+down sink, which uses a camera to find a person's head to adjust its height (Bonanni et al., 2005).
- Context-aware doll, which can emit different sounds and music according to its situation and how it is handled; it uses a combination of 16 built-in sensors including touch sensors, bend sensors,

[3] http://web.media.mit.edu/~jpatten/sensetable.html.

[4] The Sentient Materials Group is http://smg.dcl.info.waseda.ac.jp/website/home.php.

[5] http://mediacup.teco.edu/.

[6] http://www.media.mit.edu/ci/projects/chameleonmug.html.

[7] http://www.media.mit.edu/ci/projects/intelligentspoon.html.

a camera, a microphone, an accelerometer, and two infrared proximity sensors (Yonezawa et al., 2001).

■ The context-aware camera (Hakansson et al., 2003),[8] which can sense sounds, pollution in the air, temperature and smell, and create visual effects in photographs given its context.

■ SensVest (Knight et al., 2005), worn during different sporting activities has sensors to measure physiological data such as movement (via accelerometers), energy expenditure, heart rate (using electrodes to detect electrical signals that stimulate heart beats), and temperature.

■ Context-aware mobile phones.

There has been much work on mobile phones with context-aware capabilities, including self-supported and other infrastructure-based implementations. Self-supported context-aware phones include:

■ Sensay (Siewiorek et al., 2003), which, based on sensory information, can automatically perform operations on the phone such as the incoming call alert mode, send an SMS to the caller, or provide access to the electronic calendar.

■ The TEA project (Gellersen et al., 2002),[9] which attaches a hardware plug-in comprising sensors such as light sensors, microphones, an accelerometer, a skin conductance sensor, and a temperature sensor, and can detect situations of the phone such as "in hand," "on table," "in pocket," and "outdoors."

■ Work by Nokia Research (Himber et al., 2001) which studied time-series segmentation of sensor data collected using a sensor box in a mobile phone to recognize situations such as "user sits," "device is on a table," "user stands up and starts to walk," "user walks in a corridor," and "user walks outside."

■ Tuulari's Sensor Box (2000) — a prototype system using a sensor box containing sensors for acceleration, temperature, humidity, light, and conductance to implement rule-based behaviors such as "if the phone rings and it is picked up, it should stop ringing."

■ The MIThril Context-Aware Cell Phone Project,[10] which uses sensors such as a GPS receiver, an accelerometer, IR tagging, and microphone to help determine the location and activity of the user.

■ ContextPhone (Raento et al., 2005), a software platform comprising C++ libraries that implements context-aware capabilities for phones

---

[8] http://www.viktoria.se/fal/projects/photo/context.html. [9] http://www.teco.edu/tea/.

[10] http://www.media.mit.edu/wearables/mithril/phone.html.

using Symbian OS and the Nokia Series 60 Smartphone platform. ContextPhone comprises modules for sensing (supporting location information as returned by GSM cell IDs, phone information such as charger status and alarm profile, communication behavior such as calls, call attempts and SMS content, and optical marker recognition using a built-in phone camera), communications (using Generalized Packet Radio Service (GPRS), Bluetooth networking, SMS or MMS messaging), customized applications, and system services (e.g., error logging and recovery).

Specialized devices have also been developed, such as the "Hello.Wall" of the Ambient Agoras project (Prante et al., 2004), which displays different patterns to communicate messages to different audiences according to different situations of the audience.

Smart clothing is also an active area of research, including wearable computers embedded in clothing, and specialized clothing such as a raincoat that responds to rain, a dress with panels that rearrange according to time. Gershenfeld (1999) talks about smart artifacts including smart shoes, and research by the Things That Think Consortium.[11]

Skramstad (2002) defines the notion of augmented objects as "appliances whose meaning or functionality is changed or enhanced through computational power." Examples given include a café table that displays community content appropriate to the café it is located in and lamps that change appearance according to climate and traffic within the building.

Everyday objects can be augmented with context awareness, and, perhaps, new kinds of artifacts will have everyday use, artifacts that we have not yet seen or conceived; just as the era before the car, this is now the era before X, where X is still unknown.

There could be different implementations of the same concept of a context-aware artifact; i.e., the same artifact can be context-aware in different ways, different kinds of contextual information can be used, or the same contextual information can be acquired in different ways. Also, larger objects such as cars can be equipped with sensors (e.g., Vidales and Stajano, 2002) — as is increasingly done in modern cars, with light sensors for headlights and rain sensing for wipers.

Sensors added to artifacts aid in obtaining information about the situation the artifact is in, in the sense of perceiving situations from the point of view of the artifacts themselves. For example, a light sensor attached to a cup detects light according to different situations in which the cup is used. Similarly, a touch sensor attached to a soccer ball will be able to sense impacts on the ball from the ball's perspective. Hence,

---

[11] http://ttt.media.mit.edu.

such sensors attached to self-supported context-aware artifacts can acquire some kind of contextual information, not otherwise obtainable.

However, there are limitations in the number and type of sensors that can be attached to an artifact, and the use and design of the artifact itself might not facilitate such endowment of sensors to acquire needed contextual information. Moreover, artifacts might have very limited computational and networking capabilities; this limits the extent of reasoning with sensory information and prevents the artifact from connecting to Internet resources to acquire information (e.g., getting the weather report).

A different approach utilizes a hardware and software infrastructure external to the artifact. Such infrastructure-supported context-aware artifacts are liberated from what can be done with the artifact itself. To recognize the situations that a cup can be in, a whole collection of sensors can be used, and complex reasoning can be achieved with an infrastructure that "observes" the cup and its situations from "outside" the cup. The idea shifts from the artifact perceiving situations to an external party perceiving situations related to the artifact. However, a combination of sensors on the artifact and the external infrastructure can be employed. The infrastructure-supported approaches help in the sharing and management of contextual information for different context-aware artifacts and applications.

Hong and Landay (2001) defined infrastructure as "a well-established, pervasive, reliable, and publicly accessible set of technologies that act as a foundation for other systems." According to Langheinrich et al. (2000), "an infrastructure for smart things should not only consist of an architecture to represent objects and events, but also provide various *services*. Smart things (or their virtual proxies) may need location information; they want to discover services in their physical proximity, and they may want to communicate to other (possibly remote) physical objects." Sensing, reasoning, and acting can be distributed between the artifact and infrastructure.

The AmI infrastructure (Anastasopoulos et al., 2005) that is being developed was illustrated for implementing an intelligent refrigerator. This refrigerator provides services for needy persons as well as care personnel, a key feature of the refrigerator being sensors for monitoring the door state (open or closed), the temperature within the refrigerator, the content state (i.e., whether food items have expired), and whether food items are inside or outside the refrigerator (done by RFID tagging of items and checking their presence using RFID readers). Software and underlying networking infrastructure can connect a mobile device with the refrigerator sensors, so that alerts concerning the monitored events can be issued visually or sounds generated, or even sent to the mobile device (e.g., food has expired, or the refrigerator door is left open).

Another approach, whether an infrastructure is used or not, is to consider collections of context-aware artifacts (e.g., Kameas et al. (2003) and Kumar et al. (2003)). Such collaboration between smart objects can be very useful as one can piece together context as perceived by each of the artifacts, and such integrated information can provide a bigger picture of the situation related to several artifacts. For example, media-cups and doorplate sensors can be linked to infer a meeting (Gellersen et al., 2002). Your sunglasses can call out to you if they notice that you have not taken them along, given that they know the latest weather report, and also inform the cupboard to remind you to bring along suntan lotion, given that they know your possible destination. Personal Area Networks (or PANs, for short) have been proposed as a means to connect a collection of devices (using Bluetooth as the underlying technology). Such collections may be artifacts worn on a user, contained in a wallet, collocated on a table, contained in a cupboard, distributed in the home, situated in the living room, or hung on the same wall. One can imagine interesting possibilities for collections of context-aware artifacts. Siege-mund (2004) discusses cooperative smart objects and artifacts in depth, proposing a distributed tuple space model for sharing context information, formed by computational nodes located in a number of smart objects. Rules can be written to retrieve information from the distributed tuple space or to add information. Bluetooth is one possible technology for networking the nodes, supporting the tuple space abstraction.

The smart virtual counterpart (Dubendorfer, 2001) is a paradigm for associating physical real-world objects with virtual software objects. Each physical object can be associated with a virtual object or a collection of objects (e.g., a pack of compact disks) can be associated with a virtual meta object. The association can be done via RFID sensors and by addressing each physical object or collection of objects via a unique ID. The virtual object can receive, process, and store events related to the object (e.g., an object coming within range of the object — called the *sighting* of the object — or the object's going outside the range of a sensor), including a history of its uses, and can interact with other virtual objects. It is useful to quote Dubendorfer:

> A book might inform us when a new edition has been published, show the latest list of errata to us when we hold it close to a screen and suggest further reading. It could also tell how long we have spent reading it and which other books have been standing close to it on the shelf ... A personal toy could tell us about when it was on holidays with us, which hotel we visited, and which people we met there.

Table 4.1 Possible Distributions of Functionality for a Context-Aware Artifact

Architecture	Sensing	Reasoning	Initiative for Actions
1	On artifact	On artifact	From artifact
2	On artifact	On artifact	From infrastructure
3	On artifact	On infrastructure	From artifact
4	On artifact	On infrastructure	From infrastructure
5	On infrastructure	On artifact	From artifact
6	On infrastructure	On artifact	From infrastructure
7	On infrastructure	On infrastructure	From artifact
8	On infrastructure	On infrastructure	From infrastructure

4.2 ARCHITECTURAL DESIGN SPACE FOR A CONTEXT-AWARE ARTIFACT

Given an artifact, and considering that a context-aware capability is generally implemented via the three stages of sensing, reasoning about sensed information, and then initiating actions based on the reasoning, we can map out the different ways to implement context awareness for the artifact. A simple analysis of possible combinations is shown in Table 4.1, which summarizes eight possible architectures for adding context awareness to the artifact, depending on where the main processing for each stage might be physically located, either on (or augmenting) the artifact or external to the artifact, i.e., considered to be on the infrastructure side (e.g., running in a computer separate from the artifact). Architecture 1 is typical of the self-supported approach in which all three stages are done on the artifact or in hardware (and software) augmenting the artifact. Architecture 8 is typical of the infrastructure-supported approaches in which all three stages are performed on the infrastructure side. For example, the infrastructure employs a camera (considered to be on the infrastructure side) to determine the position of an object (with functionality exposed as Web services) and other entities near it, and then, according to the location of the artifact and what is nearby, makes Web service calls to adjust the functionality of the object.

Architectures 2 to 7 involve a mix of infrastructure-supported and self-supported artifact capabilities. Architecture 2 would perform sensing and reasoning on the artifact and then pass the results to the infrastructure to decide on what and how to act. This is useful if the infrastructure is coordinating a collective response of several artifacts, based on the reasoned information assembled from different artifacts. Architecture 3 would involve sensing on the artifact, passing the sensed data to the infrastructure for processing and reasoning, and then handing the results back to the artifact, which then decides and initiates appropriate actions. This is helpful if a powerful reasoning engine is used to interpret sensed data, perhaps in combination with other knowledge bases, and shared among artifacts, thereby relieving the artifact from having to be augmented with powerful computational resources. However, the artifact retains the responsibility to decide what to do, using the reasoned information fed to it by the infrastructure. Architecture 4 would employ sensors on the artifact and pass the sensor data to the infrastructure to reason and act upon. Compared to Architecture 3, the artifact relinquishes its responsibility to the infrastructure to decide and initiate actions. Architecture 5 would perform sensing via the infrastructure using sensors external to the artifact and then pass that sensor data to the artifact, which then reasons with it and finally decides what to do. This style is useful for an artifact with little or no sensing capabilities itself and is fed sensor information by the infrastructure. Architecture 6 would allow only the reasoning to occur on the artifact, which is useful to enable artifacts with their own reasoning capabilities; in this case, only reasoning is distributed to the artifacts. Architecture 7 would only inquire of the artifact what to do based on sensed and reasoned information, passing to the artifact the responsibility to decide what to do.

A distribution of functionality is also possible, for example, where sensing is done partly on the artifact and partly via the infrastructure; i.e., the artifact might decide what to do via its own sensors and via sensed information from other sensors connected to the infrastructure. Also, the artifact might negotiate with the infrastructure on what is the best course of action. We have only considered one artifact in the aforementioned analysis. In a space with multiple artifacts of different capabilities, different architectures might be employed for context-aware capabilities of different artifacts.

The sophistication of the context-aware behavior and responses would depend on the sophistication of the sensing technology used, the reasoning employed, and the actions permitted. These are not independent; for example, the reasoning component will need to be modified if new sensors that provide new kinds of sensed information are added.

A number of infrastructures and toolkits for context-aware computing, which can be used for different applications, have been developed, including those by Biegel and Cahill (2004), Henricksen and Indulska (2004), and Dey et al. (2001), as well as the ContextBroker (Chen et al., 2004). Others have been mentioned in Chapter 2. These infrastructures can be used with artifacts to endow them with context-aware behavior.

4.3 CONTEXT-AWARE MOBILE PHONES: AN ILLUSTRATION

We consider the concept of context-aware mobile phones in more detail in this section, because mobile phones are widespread and there has been much activity in this area. We have already seen examples of self-supported context-aware mobile phones. Other work such as the Situ-AwarePhone (Wang et al., 2004) uses an ontology-based approach to recognize and reason with context, using a supporting infrastructure. The context-aware phone in Connelly and Khalil (2004) utilizes an infrastructure representing the *space* in which the phone is situated, and considers how the space and phone would negotiate in determining suitable actions on the phone, such as setting the phone to silent mode in the context of a lecture. In the following detailed framework for context-aware mobile phones, we explore further this notion of space control over the phone, whether for compliance reasons (where the phone should abide by some policy concerning phone usage and behavior) or for convenience reasons (where the actions can be carried out on the phone automatically, reducing user intervention).

4.3.1 Overview of a Framework for Context-Aware Mobile Phone with User Preferences: The CAMP-UP System

CAMP-UP is an example of a system for context-aware control of mobile phones in which the level of access (from external parties) to phone functionality can be adjusted, based on the context and user preferences.

The architecture of the system comprises a server side and a client side (user side), and two classes of users are considered: the device (mobile phone) user and the space administrator. The concept of space and device follows Connelly and Khalil's design (2004). Correspondingly, there are two types of interactions we must consider as depicted in Figure 4.1:

- Interaction between the user and his or her mobile phone, where the user sets his or her preferences for the device. The user can set via his or her device the type of functionalities (e.g., vibration and phone power) that the user wants exposed and to be controllable by external parties when in different contexts. User preferences are recorded in a database on the device.
- Interaction between the space administrator and the Space Manager component of the system. The space administrator defines a space policy, which is a set of rules that determines certain device functionalities to be set to certain modes based on current context. Space policies are added to a database on the space server.

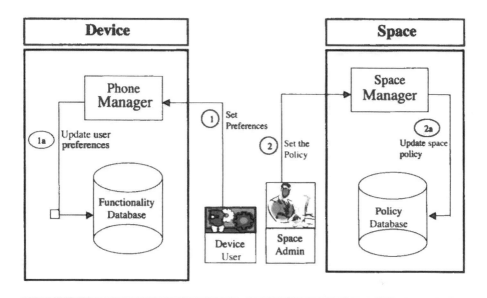

Figure 4.1   User interactions for the CAMP-UP system.

4.3.2   CAMP-UP System Interaction

The CAMP-UP system comprises two essential entities: space and device (mobile phone). Figure 4.2 illustrates the CAMP-UP system architecture. The figure uses arrows to represent the actions by the source component on the destination component. There are 14 steps of interactions. We provide an overview as follows.

In the first step, the Space Manager identifies that a device has entered the space, and the space sends context information, such as a label for the space's location and a description of the space's current activity, to the device. At the same time, the space also asks the device for the set of functionalities that it is exposing to the space, given the context information. A device is said to expose a functionality (e.g., ringer volume, vibration) if it allows the SpaceManager to have control over that particular functionality (e.g., allows the space to set the ringer volume to certain level). In Step 2, the Phone Manager passes the context information it received from the space to the Context Aggregator. The context database, in which all the context information is stored, is also updated (Step 3). Context information is also passed to the Context Interpreter to derive higher-level context information, which is also then stored in the context database. When the derived context information has been inferred and updated in the database, the Phone Manager in Step 6 gets the derived context information from the database.

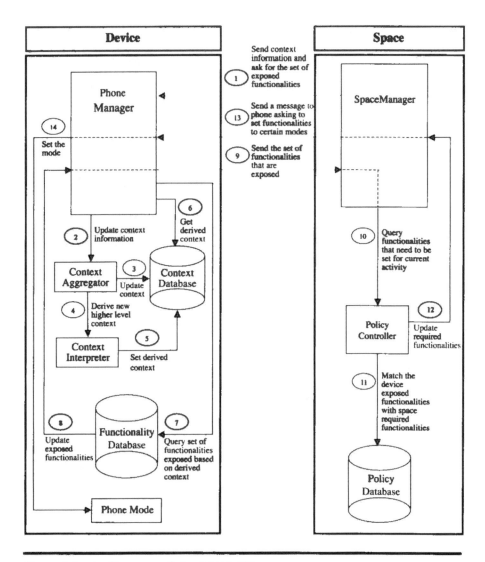

Figure 4.2   Architecture of the CAMP-UP system.

In Step 7, the Phone Manager uses the derived context information as parameter in a query to retrieve the set of exposed functionalities from the functionality database, and the response is obtained in Step 8. This response is sent back to the Space Manager as a response to the Space Manager's query in Step 1. Subsequently, in Step 10, the Space Manager determines the set of functionalities that need to be set for the current space's activity, by consulting the space policy. In Step 11, the Policy Controller component tries to match the device's exposed functionalities with space required functionalities, based on the space's current policy.

As we discussed before, the space policy is a set of rules that determines what device functionalities are to be set to what modes, based on the space's current context. In Step 12, the Policy Controller updates the Space Manager with the required functionalities. Consequently, in the next step, the SpaceManager sends a message to the phone asking to set functionalities to certain modes. In the last step, the phone manager sets the phone to certain modes, based on the space requirements.

4.3.2.1 On the Space Side

Space Manager is the main software component that is responsible for discovering new devices that enter the space and devices that leave the space. When the device enters the space, the Space Manager will initiate a new connection with the client and send the current context information to that particular client.

There are two basic types of context information that the Space Manager would send to its client:

- Location
- Activity

Location is the required context information that the Space Manager has to send to the device to permit the CAMP-UP architecture to work. Context location means the current location of the Space Manager, which is fixed for every Space Manager (e.g., home, car, room). The second type of context is the activity context that represents the current actual activity within the space. There are two different ways to attain the space's activity context. The first (preferable) method will be the current activity derived from an activity schedule. The scheduled activity will become the space's current activity if the time of scheduled activity matches with the current time. The second method is used when the space has its own context aggregator and context interpreter, as well as sensors to determine the current activity occurring within the space.

Apart from sending the context information, the Space Manager will ask for the set of exposed functionalities from any device that is currently in the space area and connected to the Space Manager. This set of functionalities will be those the Space Manager will have control over.

Space control over device functionality can be viewed from two different perspectives. The first is space control of device functionality for compliance reasons. This is when the device has no right not to follow the space policy. A scenario, for example, will be when the device is in the hospital or air plane where all the devices have to be turned off for safety purposes. The second is space control of device functionality for the convenience of the user. This is when the device has the right not to comply with the space policy. For example, when the user is watching a movie in a cinema, the device has the right not to follow the space policy, which is to silence the mobile phone, but it is convenient for the phone to be automatically set to the silent mode. Following receiving the set of functionalities from the device, the Space Manager will send a message to the device asking it to set the functionalities to prescribed modes based on the space policy.

Policy controller is a software component whose main purpose is to perform a comparison to match the set of exposed functionalities sent by the device with the space-required functionality based on the space's current policy. For example, the device sends a set of exposed functionalities, which includes ringer, vibration, and Internet connection, and the space's current policy requires control over the ringer and Internet connection; this case is a match. But it is possible that the device might send a set of functionalities that do not match with the space's current policy.

The policy database is a storage component that stores the space policy for each particular type of space activity. The space policy identifies the required settings on the device functionalities based on the space's current activity. For example, we can have a rule such as "when the space's activity context is lecture, every mobile phone is required to be silent and set to vibration off mode."

4.3.2.2 On the Client Side

To support context awareness and user preferences, the mobile phone has five components:

1. Phone Manager
2. Context Aggregator
3. Context Information
4. Context Interpreter
5. Functionality Database

Similar to the Space Manager, the Phone Manager is the main software component that controls the mobile phone interaction with the Space Manager. The four tasks of the Phone Manager are as follows:

1. Establish and maintain the connection with the Space Manager
2. Update the context aggregator whenever there is new context information received from the space
3. Send the set list of functionalities that are exposed from the phone to the space, given the context information from the space

4. Update the phone mode when the Space Manager sends a message requiring functionalities to be set to a certain mode based on the space policy

The Context Aggregator is a software component that gathers the context information. Apart from context information from the Space Manager, the context aggregator may gather context information from other sources as well, such as the device's calendar that resides in the phone itself or other sensors attached to the phone. Every piece of context information that the context aggregator attains is recorded in the Context Database. Whenever there is new context information or a context change, the context aggregator asks the context interpreter to derive new high-level contexts.

The information stored in the Context Database can be low level context information, such as location (e.g., bathroom, hotel, and restaurant) or space's current activity (lecture, meeting and conference, dinner). In addition, the information can be at a higher level, such as "lecture at Monash University" and "meeting with supervisor," derived using the inference engine, which is the Context Interpreter. The Context Database information is updated whenever the context aggregator receives new context information. For example, when the device enters the conference room, the Space Manager sends new scheduled information about the space's current activity (e.g., board meeting). The Context Aggregator then updates the Context Database with the new context information about the user and space.

The functionality database is another storage component on the client side system. The functionality database stores data of associates' phone functionalities for each *level of exposure* and rules that map situations (or derived context in this case) to levels of exposure.

The ability to programmatically change the modes on a phone depends on the application programming interface (API) available for the phone. For instance, the Java Telephone API (JTAPI),[12] is a variation of the PersonalJava application environment, and it is a J2ME (Java 2 Microedition) API for the design, development, and deployment of MIDP Java applications for cell phones, including Wireless Application Protocol (WAP) and i-Mode. JTAPI provides a number of methods that could be used to adjust the functionalities' settings, such as the setRingerVolume() method in the PhoneRinger class, which would set the ringer volume of the phone. There are a number of other phone APIs available based on the phone manufacturer and model. For example, the Nokia 3410 SMS API enables sending and receiving of GSM short messages from and to J2ME MIDP applications. The methods we could use are the send(Message msg) and receive() in the MessageConnection class. The API is based on the Generic

---

[12] http://java.sun.com/products/jtapi/.

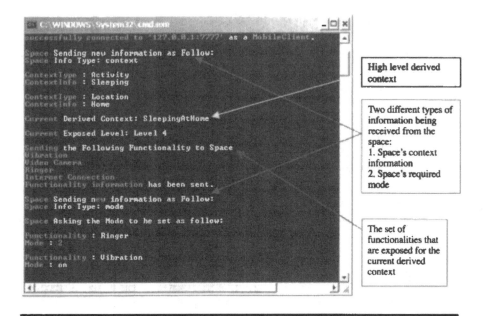

Figure 4.3 Device send and receive information.

Connection Framework (javax.microedition.io package) that is used for I/O and networking functionality in the J2ME profiles. In the future, there could emerge a comprehensive API, enabling automatic control over phone functionalities — such an API can be a concern from the security perspective, and so, its use requires careful regulation (technical means or otherwise).

4.3.3 Prototype and Discussions

The CAMP-UP system was implemented using a PocketPC (simulating a mobile phone) and a desktop PC (simulating the space server). Figure 4.3 shows text debug messages from the prototype system. The screen shot shows context information being sent to the device, followed by a return in the level of exposure to the space (which in this case is level 4), and the space, in turn, setting the ringer volume to 2 and turning the vibration mode on.

The derived context shown in Figure 4.3 is deduced by using simple rules such as the following (more sophisticated reasoning can alternatively be employed):

```
IF (home) THEN
    IF (sleeping) THEN
        setDeriveContext (sleepingAtHome)
```

```
IF (studying) THEN
    setDeriveContext (studyingAtHome)
IF (showering) THEN
    setDeriveContext (showeringAtHome)
```

The interface for the client side of the CAMP-UP system has two main features in its main menu. The first one is the feature to set the level of exposure for different types of context. The second is to allow the user to associate different phone functionalities for different levels of phone exposure. Figure 4.4 shows the interface for the two main client features.

As illustrated in Figure 4.5, each recognized context in the CAMP-UP system is mapped to a different level of exposure. For example, home context has exposure level 5 and sleeping at home context has exposure level 4. This means that the device will expose the functionalities (e.g., power, vibration, etc.) that are included in level 4 to the home Space-Manager if given the sleeping-at-home context.

With the current client interface, there are seven levels of functionality exposure. The number of default levels depends on the number of different functionalities that the phone has, and the levels could correspond to levels of obtrusiveness to the user so that the user associates how obtrusive the phone should be in different situations. The user will only have the capability to view the functionalities that are exposed for a given level of exposure as shown in Figure 4.6.

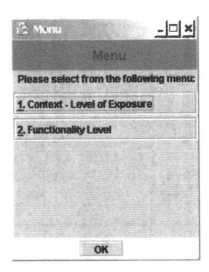

Figure 4.4   CAMP-UP client main menu.

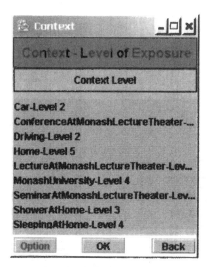

Figure 4.5   Context — level of exposure.

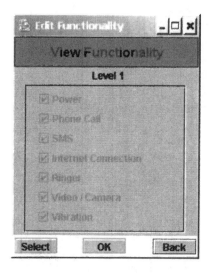

Figure 4.6   Exposed functionalities for level-1 exposure.

Table 4.2 A Summary of Scenarios Discussed

Scenario	Number of Exposed Functionalities	Number of Required Functionalities	Number of Matched Functionalities	Compliance or Convenience
1	2	2	2	Convenience
2	7	1	1	Compliance
3	1	2	1	Convenience
4	1	2	1	Compliance
5	1	2	0	Convenience
6	6	1	0	Compliance

4.3.4 Scenarios

A variety of scenarios are used to illustrate the different situations that a mobile phone might encounter in the real world. Table 4.2 summarizes six scenarios, in terms of the number of exposed functionalities, the functionalities required of the space, the number of matched functionalities between the exposed and required functionalities, and the main purpose of space control (compliance or convenience as explained later).

Scenario 1: User Attending Lecture in Monash Lecture Hall

A student of Monash University is attending a lecture in the lecture hall.

Space Policy

Monash University advises that all the mobile phones should be switched to silent mode during the lecture and tutorial time.

User Preference

The student would like a customized level of exposure. Based on user preferences, that is "customized 3," there are two functionalities exposed in this context, which are the ringer and vibration settings as shown in Figure 4.7.

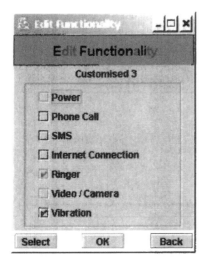 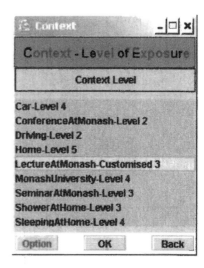

Figure 4.7   The result of device interaction in this scenario.

Figure 4.8   Scenario 1 debug messages from the space's perspective.

As shown in Figure 4.8, there are two required functionalities for the current space policy. The figure also shows that there are two matched functionalities. The space then sends the message asking the phone to set the ringer volume to level 0 and vibration to the off mode. Referring to Figure 4.9, we could see the message from the space asking the device to set the functionalities to certain modes.

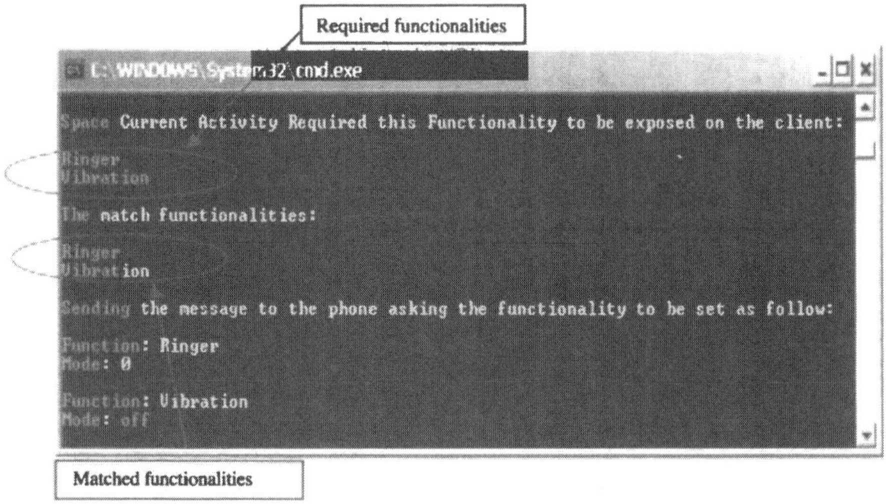

Required functionalities

Matched functionalities

Figure 4.9 Scenario 1 debug messages from the device's perspective.

Scenario 2: User in the Emergency Room of the Hospital

User accompanies a relative to an operation room of the hospital.

Space Policy

The hospital strictly requires all mobile phones to be switched off in the emergency and operation rooms to avoid interference caused by mobile phone signals with medical equipment.

User Preference

The user prefers to give full authority to the hospital Space-Manager over the user's phone. This is expressed as exposing functionality Level 1 to the hospital SpaceManager. Based on user preferences defining Level-1 exposure, all functionalities are exposed to be controllable by the space. As the space has received full control over the phone, it is authorized to ask the phone to turn its power off.

Scenario 3: User Sleeping at Home

On Saturday afternoon, the user is taking a nap at home.

Space Policy

Home SpaceManager would like the phone to be in the silent mode and vibration to be off so that the incoming call will not disturb the user.

User Preference

The user would like the phone to expose functionality level 7, which means that the vibration functionality is exposed. The reason for this is that the user would like the phone ringer volume to stay on the current setting as, while he is sleeping, he is expecting an important call from a partner. There are two functionalities that are required from the device based on the current space policy. However, only one functionality is exposed, the vibration settings; i.e., in this case, the user's preferences are respected. Although the space wants to control the ringer volume as well, the phone only allows the vibration setting to be controlled by the space. Therefore, the space can only ask the phone to set the vibration mode off.

Scenario 4: User Driving, in the Car

User is driving to office in a car.

Space Policy

For road safety purposes, all drivers are strictly banned from using cell phones in cars. Therefore, the car space policy requires the cell phone to be in the silent mode and the vibration mode to be off. Based on this policy, the SpaceManager of the car requires control of the ringer and vibration functionalities from the device.

User Preference

The user has made a customized level of exposed functionality for the driving context that only allows the ringer volume to be controlled by the car's SpaceManager. There are two required functionalities for the current space policy. However, only the vibration functionality is exposed.

Scenario 5: User Watching a Movie in a Cinema

The user is watching a movie in a cinema.

Space Policy

The cinema urges customers to put their phones to silent mode during the movie, so that the phone will not distract other customers.

User Preference

The user enables the phone to be exposed to the cinema's SpaceManager, but, based on the user's preference for this particular situation, the phone exposes only the SMS functionality. As the current space policy requires the phone to expose two functionalities, the ringer and the vibration, there is no matching functionality because of the user's preference. But the SpaceManager could send an SMS message to the phone asking the user to set the functionalities based on the space policy.

Scenario 6: User on Board an Airplane

User is on the plane.

Space Policy

Mobile phone is strictly required to be switched off during the flight to avoid interference with airplane equipment.

User Preference

The user enables Level-2 exposure to the plane's SpaceManager, which includes all the device's functionalities, except power. Because of the space policy requiring the mobile phone to be switched off and the current user's preference not enabling this phone functionality to be exposed, the SpaceManager cannot turn the phone off automatically.

We have seen that in some scenarios, there might be a case in which the space policy requires functionalities not fully match-

ing the device's exposed functionalities as shown in scenarios 3 and 4, in which the space required two functionalities whereas only one functionality is exposed. In addition, there are instances in which no functionality from the device matches the space policy requirement, as shown in scenarios 5 and 6.

Space control over device functionality can be viewed from two different perspectives. The first is the space control of device functionality for compliance reasons; this means that the device has an obligation to follow the space's policy. Scenarios 2, 4, and 6 illustrate the compliance view, where the SpaceManager has strict requirements about the functionalities it needs to control and, therefore, about the functionalities the phone should expose.

In Scenario 2, the SpaceManager that belongs to the hospital strictly requires all mobile phones to be switched off when the user is in the emergency or operation room to avoid the interference of mobile phone signals with medical equipment. A similar case for the car SpaceManager is Scenario 4, in which the user (or device) has to comply with the road safety rule that requires all drivers not to use their mobile phones while driving. In addition, Scenario 6 illustrates the airplane SpaceManager strictly requiring all electronic equipment, including mobile phones, to be turned off during flight to avoid interference with airplane equipment.

There are a number of ways to enforce compliance by the user's device, such as asking the user to sign an agreement before utilizing services or coming into particular spaces. For compliance, there has to be a mechanism to ensure that the device obeys the space requirements. The implementation has tried to address this issue by having a default exposure level that has a predetermined set of functionalities exposed, and such exposure levels can be associated with particular situation types as a default setting on the device. However, in the current prototype implementation, there is no mechanism to enforce this — the user might associate a level of exposure to a situation type, different from the default level and, thus, not comply with the space policy.

The second perspective is the space control of device functionality for the convenience of the user; i.e., the space automatically sets the phone to particular modes to avoid the users having to do this manually. This perspective is illustrated in scenarios 1, 3, and 5.

One can think of many situations in which the phone can find itself, such as in a restaurant, in a bus, or in a meeting, where appropriate context-aware actions can be taken, whether for compliance or convenience reasons. Such context-aware behavior of the phone might also be employed beyond compliance and user convenience reasons, but for specific applications, such as changing the phone's wallpaper (containing some suitable advertisements) when passing through different spaces (say, along a street). Automated negotiation between the space and device involving software agents might be employed to determine agreed levels of control, but, where compliance is involved, enforcing a policy, whether utilizing an incentive or penalty mechanism, would involve issues beyond a mere technical solution.

4.4 SUMMARY

We have reviewed context-aware artifacts, and we imagine that new artifacts are being developed and will emerge in the coming years. Developing such artifacts involves a combination of engineering and computing expertise, including hardware and software. We have presented the work in the two categories of self-supported and infrastructure-supported context-awareness, following Loke (2006). In addition, different ways in which sensing, acting, and reasoning can be distributed for developing context-aware artifacts — that is, the possible architectures — have been examined, leading to interesting design possibilities. Our example of the CAMP-UP system is detailed and illustrates an implementation of a context-aware mobile phone using the space and artifact model. Stepping back from mobile phones, the interaction between the space and an artifact can be further explored for different kinds of artifacts and different spaces; e.g., when the artifact moves into different spaces, it might be endowed with different possible sensory information and be capable of new actions, using sensors and effectors at the infrastructure or space end. For example, robots in households need not only work with the sensors on them but also utilize the sensors on the infrastructure (e.g., a positioning infrastructure) in the home.

REFERENCES

Agarawala, A., Greenberg, S., and Ho, G., The Context-Aware Pill Bottle and Medication Monitor, Technical Report 2004-752-17, Department of Computer Science, University of Calgary, Calgary, Alberta Canada.

Anastasopoulos, M., Bartelt, C., Koch, J., Niebuhr, D., and Rausch, A., Towards a reference middleware architecture for ambient intelligence systems, *Proceedings of the Workshop on Building Software for Pervasive Computing at OOPSLA*, 2005, available at http://www.ics.uci.edu/~lopes/bspc05/papers/anastopoulos.pdf.

Beigl, M., Gellersen, H-W., and Schmidt, A., MediaCups: experience with design and use of computer-augmented everyday objects, *Computer Networks*, special issue on Pervasive Computing, Elsevier, 2001.

Biegel, G. and Cahill, V., A framework for developing mobile, context-aware applications, *Proceedings of the 2nd IEEE Conference on Pervasive Computing and Communications*, Percom 2004, Orlando, FL, March 14–17, to appear, available at Department of Computer Science, Trinity College Dublin, Technical Report TCD-CS-2004-04, available at http://www.cs.tcd.ie/publications/tech-reports/reports.04/TCD-CS-2004-04.pdf.

Bonanni, L., Arroyo, E., Lee, C.-H., and Selker, T., Smart sinks: real-world opportunities for context-aware interaction, *Proceedings of the CHI 2005*, Portland, OR, April 2005, ACM Press, 1232–1235, available at http://web.media.mit.edu/~amerigo/p1232-bonanni.pdf.

Brooks, R.A., A Robust Layered Control System for a Mobile Robot. A.I. Memo 864, MIT, U.S.A., September 1985.

Chang, K.-H., Liu, S.-Y., Chu, H.-H., Hsu, J., Chen, C., Lin, T.-Y., Chen, C.-Y., and Huang, P., Dietary-aware dining table — observing dietary behaviors over tabletop surface, *Proceedings of the 4th International Conference on Pervasive Computing (Pervasive 2006)*, Dublin, Ireland, May 2006, available at http://mll.csie.ntu.edu.tw/papers/diettable_pervasive2006.pdf.

Chen, H., Finin, T., and Joshi, A., A context broker for building smart meeting rooms, *Proceedings of the Knowledge Representation and Ontology for Autonomous Systems Symposium*, C. Schlenoff and M. Uschold, Eds., 2004 AAAI Spring Symposium, Stanford, CA, 2004, AAAI Press, pp. 53–60.

Christensen, H.I., Intelligent home appliances, in *Robotics Research*, Jarvis, R.A., and Zelinsky, A., Eds., No. 6 in Springer Tracts in Advanced Robotics (STAR), Springer-Verlag, 2003, pp. 319–330.

Connelly, K. and Khalil, A., On negotiating automatic device configuration in smart environments, *Proceedings of PerWare 04 Workshop, 2nd IEEE International Conference on Pervasive Computing and Communications*, Orlando, FL, March 14–17, 2004.

Dey, A.K., Salber, D., and Abowd, G.D., A conceptual framework and a toolkit for supporting the rapid prototyping of context-aware applications, in special issue on context-aware computing in the *Human-Computer Interaction (HCI) Journal*, 16(2–4), 97–166, 2001.

Dubendorfer, T., An Extensible Infrastructure and a Representation Scheme for Distributed Smart Proxies of Real World Objects, Master's thesis, Swiss Federal Institute of Technology, Zurich, April 2001, available at http://www.vs.inf.ethz.ch/publ/papers/TR_359.pdf.

Fujinami, K., Kawsar, F., and Nakajima, T., Aware mirror: a personalized display using a mirror, *Proceedings of the International Conference on Pervasive Computing (Pervasive 2005)*, Munchen, Deutschland, May 2005.

Gatenby, D.A.G., Galatea: Personalized Interaction with Augmented Objects, Master of Science thesis, MIT, September 2005, available at http://web.media.mit.edu/~dagg/thesis/ThesisFinal.pdf.

Gellersen, H.W., Schmidt, A., and Beigl, M., Multi-sensor context-awareness in mobile devices and smart artifacts, *Mobile Networks and Applications* (MONET), October 2002, available at http://www.comp.lancs.ac.uk/~hwg/publ/monet.pdf.

Gershenfeld, N., *When Things Start to Think*, Henry Holt and Company, U.S.A., 1999.

Hakansson, M., Ljungblad, S., and Holmquist, L.E., Capturing the invisible: designing context aware photography, *Proceedings of DUX 2003, Designing for User Experience*, ACM Press, available at http://www.viktoria.se/fal/projects/photo/DUX_hakansson.pdf.

Headon, R., Movement awareness for a sentient environment, *Proceedings of the 1st Conference on Pervasive Computing and Communications (PerCom 2003)*, March 2003, IEEE Computer Society Press.

Henricksen, K. and Indulska, J., A software engineering framework for context-aware pervasive computing, *Proceedings of the 2nd Conference on Pervasive Computing and Communications (PerCom 2004)*, 2004, pp. 77–86.

Himber, J., Tikanmäki, J., Toivonen, H.T.T., Korpiaho, K., and Mannila, H., Time series segmentation for context recognition in mobile devices, *Proceedings of the 2001 IEEE International Conference on Data Mining (ICDM'01)*, San Jose, CA, IEEE Computer Society Press, 2001, pp. 203–210, available at http://www.nokia.com/downloads/aboutnokia/research/library/software_computing/SWC5.pdf.

Hong, J.I. and Landay, J.A., An infrastructure approach to context-aware computing, *Human-Computer Interaction (HCI) Journal* 16(2–3), 2001.

Ito, M., Iwaya, A., Saito, M., Nakanishi, K., Matsumiya, K., Nakazawa, J., Nishio, N., Takashio, K., and Tokuda, H., Smart furniture: improvising ubiquitous hot-spot environment, *Proceedings of the 3rd International Workshop on Smart Appliances and Wearable Computing at ICDCS*, Providence, RI, 2003, IEEE Computer Society, pp. 248–253.

Kameas, A., Bellis, S., Mavrommati, I., Delaney, K., Colley, M., and Pounds-Cornish, A., An architecture that treats everyday objects as communicating tangible components, *Proceedings of the 1st IEEE International Conference on Pervasive Computing and Communications (PerCom'03)*, available at http://cswww.essex.ac.uk/Research/iieg/papers/egadgets-percom03.pdf.

Knight, J.F., Schwirtz, A., Psomadelis, F., Baber, C., Bristow, H.W., and Arvanitis, T.N., The design of the SensVest, *Personal and Ubiquitous Computing* 9, 6–19, 2005, Springer.

Kumar, R., Poladian, V., Greenberg, I., Messer, A., and Milojicic, D., Selecting devices for aggregation, *Proceedings of the IEEE Workshop on Mobile Computing Services and Applications*, 2003.

Langheinrich, M., Mattern, F., Römer, K., and Vogt, H., First steps towards an event-based infrastructure for smart things, *Proceedings of the Ubiquitous Computing Workshop (PACT 2000)*, Philadelphia, PA, October 15–19, 2000, available at http://www.vs.inf.ethz.ch/publ/papers/firststeps.pdf.

Loke, S.W., Context-aware artifacts: two development approaches, *IEEE Pervasive* 5(2), 48–53, 2006.

Loke, S.W., Syukur, E., and Stanski, P., Adding Context-Aware Behaviour to Almost Anything: the Case of Context-Aware Device Ecologies, accepted for the MobiSys 2004 workshop on context-awareness, available at http://www.sigmobile.org/mobisys/2004/context_awareness/papers/mobisys-ca.pdf.

Matthews, T., Gellersen, H-W., Van Laerhoven, K., and Dey, A.K., Augmenting collections of everyday objects: a case study of clothes hangers as an information display, *Proceedings of Pervasive'04*, 2004, pp. 340–344.

Metzger, C., Indirect object-sensing technology to prevent out-of-stocks at retail-level, *Proceedings of the Workshop on Smart Objects at the International Conference on Ubiquitous Computing (UbiComp05)*, Tokyo, Japan, 2005, available at http://ubicomp.lancs.ac.uk/workshops/sobs05/papers/Metzger,%20Christian.pdf.

Mihailidis, A., Tse, L., and Rawicz, A., A context-aware medication reminding system: preliminary design and development, *Proceedings of the Rehabilitation Engineering and Assistive Technology Society of North America*, Atlanta, GA, June 2003.

Nakajima, T., Fujinami, K., and Tokunaga, E., Building intelligent environments using smart daily objects and personal devices. *Proceedings of the Workshop on Context Awareness for Proactive Systems (CAPS2005)*, Helsinki, Finland, June 2005, available at http://www.dcl.info.waseda.ac.jp/publications/pdf/ic2005-building_intelligent.pdf.

Patten, J., Ishii, H., Hines, J., and Pangaro, G., Sensetable: a wireless object tracking platform for tangible user interfaces, *Proceedings of the Conference on CHI*, 2001, ACM Press, pp. 253–260.

Picard, R.W. and Klein, J., Computers that recognise and respond to user emotion: theoretical and practical implications, *Interacting with Computers* 14(2), 141–169, 2002.

Prante, T., Stenzel, R., Rocker, C., Streitz, N., and Magerkurth, C., Ambient agoras: InfoRiver, SIAM, Hello.Wall, *Proceedings of the Conference on Human Factors in Computing Systems (CHI'04)*, Vienna, Austria, 2004, ACM Press, pp. 763–764.

Raento, M., Oulasvirta, A., Petit, R., and Tolvonen, H., Contextphone: a prototyping platform for context-aware mobile applications, *IEEE Pervasive Computing* 4(2), 51–59, 2005.

Salvador, Z., Bonail, B., Lafuente, A., Larrea, M., Abascal, J., and Gardeazabal, L., AmIChair: ambient intelligence and intelligent wheelchairs, *Proceedings of the Home Oriented Informatics and Telematics Conference*, Vol. II, HOIT 2005, York, U.K., 2005, pp. 31–36.

Schilit, B.N., Adams, N., and Want, R., Context-aware computing application, *Proceedings of the Workshop on Mobile Computing Systems and Applications*, Santa Cruz, CA, 1994, IEEE Computer Society, pp. 85–90.

Selker, T., Arroyo, E., and Burleson, W., Chameleon tables: using context information in everyday objects, *Proceedings of the Conference on Human Factors in Computing Systems*, CHI'02 extended abstracts on human factors in computing systems, Minneapolis, MN, 2002, pp. 580–581.

Siegemund, F., Cooperating Smart Everyday Objects — Exploiting Heterogeneity and Pervasiveness in Smart Environments, Doctor of Technical Sciences Dissertation, Swiss Federal Institute of Technology (ETH Zurich), 2004.

Siewiorek, D., Smailagic, A., Furukawa, J., Krause, A., Moraveji, N., Reiger, K., Shaffer, J., and Wong, F.L., SenSay: a context-aware mobile phone, *Proceedings of the Seventh IEEE International Symposium on Wearable Computers (ISWC'03)*, available at http://csdl.computer.org/comp/proceedings/iswc/2003/2034/00/20340248.pdf.

Skramstad, H., Augmented Objects — Blending Bits and Atoms, 2002, available at http://www.ivt.ntnu.no/ipd/fag/PD9/2002/Artikler/Skramstad%20I.pdf.

Steurer, P. and Srivastava, M.B., System design of smart table, *Proceedings of the IEEE International Conference on Pervasive Computing and Communications*, Dallas-Fort Worth, TX, 2003, pp. 473–480.

Tuulari, E., Context-Aware Hand-Held Devices, Technical Report of the Technical Research Centre of Finland, VTT publications 412, 2000, available at http://www.inf.vtt.fi/pdf/publications/2000/P412.pdf.

Vidales, P. and Stajano, F., The sentient car: context-aware automotive telematic, poster in *Proceedings of Ubicomp 2002*, Goteborg, Sweden, available at http://www-lce.eng.cam.ac.uk/~pav25/publications/Ubicomp2002(poster).pdf.

Vildjiounaite, E., Malm, E.-J., Kaartinen, J., and Alahuhta, P., Context awareness of everyday objects in a household, *Proceedings of Ambient Intelligence, 1st European Symposium, EUSAI 2003*, Aarts, E.H.L., Collier, R., van Loenen, E., and de Ruyter, B.E.R., Eds., Veldhoven, The Netherlands, 2003, Lecture Notes in Computer Science 2875, pp. 177–191.

Wang, X., Zhang, D., Dong, J.S., Chin, C., and Hettiarachchi, S.R., Semantic space: a semantic web infrastructure for smart spaces, *IEEE Pervasive Computing* 3(3), 32–39, 2004.

Wooldridge, M. and Jennings, N.R., Intelligent agents: theory and practice, *The Knowledge Engineering Review* 10(2), 115–152, 1995.

Yamabe, T., Fujinami, K., and Nakajima, T., Experiences with building sentient materials using various sensors, *Proceedings of the 24th International Conference on Distributed Computing Systems Workshop (ICDSCW 2004)*, IEEE Computer Society Press.

Yonezawa, T., Clarkson, B., Yasumura, M., and Mase, K., Context-aware sensor-doll as a music expression device, *Proceedings of Conference on Computer-Human Interface*, 2001, available at http://www.mic.atr.co.jp/~yone/chi2001/.

5

CONTEXT-AWARE MOBILE SOFTWARE AGENTS FOR INTERACTION WITH WEB SERVICES IN MOBILE ENVIRONMENTS

In this chapter, we discuss intelligent software agents running on mobile agents that are context aware for interacting with Web services. This framework is called Context-Aware Lightweight Mobile BDI Agents (or CALMA, for short). We also present the implementation and evaluation of the CALMA framework. This chapter differs from Chapter 3 on context-aware mobile services in that we attempt a deeper exploration here of the notion of the agents to encapsulate and embody functionality and intelligent behavior; we think of the agents, and not the services, as being context aware. Rather than considering the context of users of mobile devices, or the context surrounding an artifact, we consider the context in which software agents find themselves.

5.1 AGENTS: MOBILE AND INTELLIGENT

Mobile agents have been explored as a technology to support mobile and ubiquitous computing because of their host-to-host migratory attribute (Kotz and Gray, 1999). The ability for computations to move from host to host on a network in an autonomous fashion becomes particularly suitable in environments that are characterized by limited and dynamically changing availability of computational resources and levels of connectivity. In this context, several mobile agent toolkits have been developed that provide runtime support for mobile agents in resource-constrained environments such as the Mobile Agent Environment (MAE) (Mihailescu and Kendall, 2002) for the Palm OS™ and Grasshopper (http://www.grasshopper.de) for WinCE™. However, the focus of mobile agent toolkits for resource-constrained mobile devices has been on the provision of support for migration. The issue of context awareness is typically delegated to the agent application. Context awareness is a key attribute for the successful deployment of mobile agents in ubiquitous environments and in fact forms part of the rationale for their usage. It is this attribute that enables an agent to perform tasks such as migrate when resources are running low, migrate when a connection is available, and avoid embarking on a task that requires continuous connectivity when the device is experiencing frequent disconnections. However, the incorporation of context awareness into mobile agents operating in resource-constrained environments brings with it a computational overhead that needs to be taken into consideration.

5.2 SCENARIOS

Web browsers mediate between users and the underlying machinery of the Web, providing a layer of abstraction and handling errors as pleasantly as possible. Web services provide a programmatic view of the Web, so that applications can harness the Web's advantages of accessibility and standardized protocols. Web services are invoked from within the context of an application, which is some program that contains Web service calls. For example, we might have an application that obtains details about user's preferences about movies and invoke appropriate Web services (or a composition of services) to book tickets for the user at specific times and cinemas. The application has to handle failures in the invocation of the services and processes the results accordingly, which can happen in the fixed network but becomes more apparent in wireless mobile computing environments. If the application is running on a mobile device, it will face the often-cited challenges with mobile computing, including disconnection handling, varying bandwidth, limited device resources (such as memory and battery power), variations in the server side (e.g., the servers of Web services are down), and changes due to mobility of the user (and device) that affect the relevance of services invoked and permit opportunistic interactions — for example, the user happens to walk near a movie theater with a direct short-range wireless connection to its system, and cost of wireless Internet connections can therefore be avoided if the short-range connection is used. In summary, it would be beneficial if that application was context aware, proactive, autonomous and mobile in its

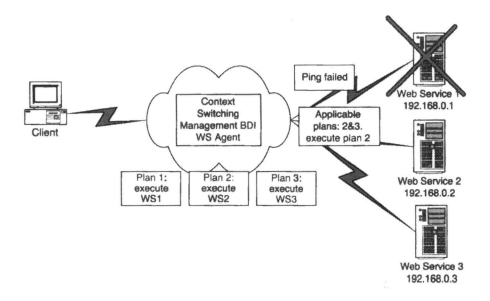

Figure 5.1   Context Switching Management Agent.

behavior, and adaptive to its environment, i.e., to both the device and the physical environment of the user. Such software properties have been discoursed in the area of intelligent software agents.

There are several scenarios that motivate the usefulness of the Context-Aware Web Services Agents that can operate from mobile devices. Consider an agent running on a machine connected to a Web farm, as shown in Figure 5.1. The Web farm consists of two servers that host the same Web services. The agent acts as a Context Switching Manager that filters the request from the client application and redirects the call to whichever server is available at that particular moment. To perform this task, the agent is fed with two different plans. Each plan redirects the Web service call by repackaging the SOAP call to a particular server. In each of the plans, a context condition that represents the availability of servers is identified by their IP addresses. The context condition in the plan will be checked before and during the execution of the plan. When a client application invokes a Web service, the agent executes its Beliefs-Desires-Intentions (BDI) logic by finding the appropriate plan from its plan library. The agent's BDI interpreter checks the context condition by sending a "ping" packet to each server. If the server responds, the plan is considered to be a valid plan and will be executed by the agent. A plan might have a series of Web service calls and act as a Web service workflow. During the execution of the Web services, the agent can detect the availability of the server on each Web service call in the plan. If one of the servers

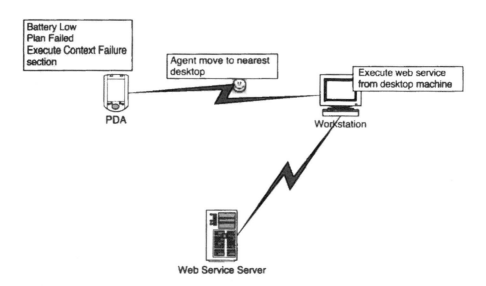

Figure 5.2 Context-Aware Web Services Mobile Agent.

is unavailable during plan execution, the agent can still complete the plan by automatically replanning itself and switching the execution to other servers.

Consider another scenario where an agent executing on a small handheld device uses Web services as shown in Figure 5.2. Before calling Web services, the handheld device must be able to detect the availability of the connection between the handheld itself and the server. The agent's plan can be programmed to allow Web service calls only when the battery power and memory resource are adequate for calling the service. Moreover, the agent can be planned to move to another machine when the battery power is low during plan execution.

5.3 A BRIEF REVIEW OF AGENT PLATFORMS FOR UBIQUITOUS COMPUTING

Before proceeding to describe CALMA in detail, in this section, we consider the following agent systems based on the review presented by Sumartono et al. (2004). AbiMA (Rahwan et al., 2003), MyCampus (Sadeh et al., 2002), Agent Factory (Collier and O'Hare, 1999), and WAY (Lowen et al., 2002) provide a mobile device execution environment for their agents. AbiMA and Agent Factory agents are not integrated with Web services, and the agents do not possess the attribute of mobility. Although MyCampus agents use Web services to access contextual data and user profiles, the agents are stationary at the server side. The integration of software agents with Web services allows them to deliver complex services to mobile users.

The system shown in Table 5.1 indicates a trend toward combining the features of reasoning, mobility, context awareness, Web service access, and support for mobile devices. Such features tend to complement each other and enable software applications with robust adaptive behavior.

5.4 CALMA ARCHITECTURE

CALMA uses the BDI agent model to build context awareness into mobile agents operating in ubiquitous environments. The BDI agent model has been used in complex multiagent systems and draws inspiration from the philosophy of human mentalistic concepts of beliefs, desires, and intentions (Wooldridge, 2002). From the programming perspective, two key concepts in the BDI model are goals and plans, the procedure for accomplishing goals. CALMA includes a plan language that is implemented in XML, which facilitates the specification of plans for mobile agents based on contextual conditions. The CALMA framework enables agents to be lightweight by using two strategies:

1. A support infrastructure that facilitates plans to be obtained "on demand" based on changing contextual conditions
2. Using the Web services paradigm to off-load computations to external servers

A unique feature of CALMA is that it is implemented as an add-on to an existing mobile agent toolkit, thereby enabling the framework to be built on existing technologies rather than redeveloping the required mobility infrastructure. Thus, CALMA functionality can be integrated with existing agent toolkits.

As shown in Figure 5.3, the infrastructure contains three components: CALMA task agent component, mobile device component, and server component. The CALMA task agent component implements the BDI agent model to provide the necessary functionality for task-specific agent implementation. The server component provides services such as matchmaking service, plan request service, and service request handling to support the mobile device component. The CALMA task agent component is representative — there can be more than one such task agent. The mobile device component contains the user interaction module to support the mobile user in requesting services and to allow the user to manage the task agents running on the device.

Table 5.1   Comparison of Related Agent Systems

System	BDI/Intelligent Behavior	Mobility	Context Awareness	Support for Semantic Web or Web Services	Support Agent Applications on Mobile and Small Devices
JAM	BDI	X	X		
AbiMA	BDI		X		X
Nuin[a]	BDI	Relies on features available on the selected underlying agent platform	X		
MyCampus	Intelligent		X	X	X
Agent Factory and WAY	BDI	X	X		X

[a] From Dickinson, I., and Wooldridge, M., Towards practical reasoning agents for the semantic Web, *Proceedings of the 2nd International Joint Conference on Autonomous Agents and Multiagent Systems*, 2003, pp. 827–834. With permission.

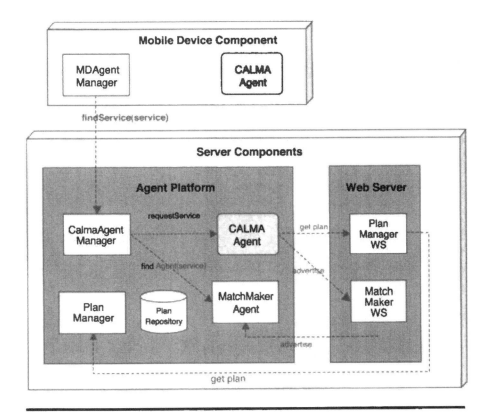

Figure 5.3 Overview of CALMA infrastructure.

5.4.1 CALMA Agent Model

The CALMA BDI framework has the following properties:

- *Rationality*: Represented by reasoning, based on the notion of BDI agent mental states (Beliefs, Desires, Intentions, and Plan Library).
- *Mobility*: The agent needs to move from server component to mobile device component, mobile device to other hosts, or vice versa. The agent performs its mobility for the following reasons:
  - Efficient localized communication with other agents that reside in different contexts.
  - The current machine context may not have enough resources to execute actions in the plan. This requires the agent to move to another machine that has more resources. For instance, the agent might prefer to perform computation on the desktop machine rather than on small devices that have fewer resources.

- The move is for an application-specific reason. This is implemented by declaring the *move* command as one of the actions in the agent's plan.
- *Context awareness*: During deliberation time, the agent needs to run on the mobile device that has limited resources. Because the resources are limited, running agents in the mobile device might experience resource problems such as low battery power or insufficient memory.
- *Social behavior*: The agent needs to communicate with another agent or service when performing its tasks.
- *Explicit Web service interactions*: This enables management of Web service interactions and invocations by the agent.

Figure 5.4 shows the CALMA BDI agent architecture. The BDI agent architecture is divided into three main parts: the *BDI* Class, the *Mobile Agent*, and *External* parts. The components that reside inside the BDI Class are the core components of the CALMA BDI agents. The Mobile Agent comes with a Mobile Agent toolkit. When the CALMA BDI agents migrate, the BDI agents carry components in both BDI Class and Mobile Agent. External components are those that reside in the mobile device, and will not be carried by the BDI agents during migration.

The CALMA BDI agent architecture consists of the following components:

- *Resource Monitor*: The Resource Monitor component is responsible for monitoring the environment conditions during the deliberation process. This component is the implementation of *context awareness* for the CALMA BDI agent. Other context information can be included, but the prototype focuses on resources. A Resource Monitor DLL component is a collection of Win32 dynamic link libraries for resource monitoring purposes. This component will be called by the Resource Monitor component. Because the Resource Monitor component is built using Java, the Resource Monitor interfaces with the Resource Monitor DLL via *Java Native Interface (JNI)*.
- *Beliefs*: The Beliefs component handles the agent's belief storage. The agent's Beliefs can be obtained from two different sources: the Interpreter (e.g., Web services response, agent communication response) and the Resource Monitor component (e.g., free memory, battery power).
- *Goals*: The Goals component keeps the set of goals. A goal is the trigger for the agent's deliberation process. The goal could represent a task submitted by users. Presently, the CALMA agent goal is top level only. In the future, a goal can consist of top-level goals

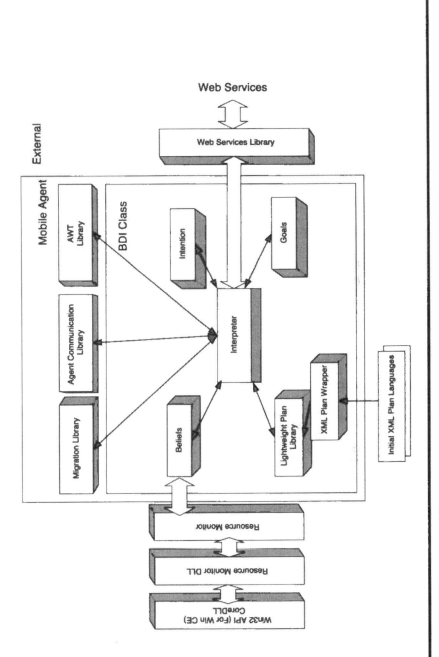

Figure 5.4 CALMA BDI agent architecture.

and subgoals. The agent might distribute the subgoals to other CALMA agents.

■ *Intention*: Intention is basically a plan that is ready to be executed. The intention consists of actions that will be sequentially executed by the Interpreter.

■ *Interpreter*: The Interpreter component is responsible for finding an intention and executing the intention. During the process of executing actions in the plan, the Interpreter manages communicating with other agents, migrating the agent to another place, displaying a graphical interface, or invoking Web services.

■ *Lightweight plan library*: A CALMA agent carries only plans related to the task that is currently processed. The agents with lightweight plans are formed by the server components. Plans can be obtained from the PlanManager when required.

■ *XML plan language wrapper*: It is responsible for wrapping XML data structure to yield an object-oriented data structure.

■ *Web services library*: This library handles Web service calls. The results from this component will be passed on to the Interpreter component.

■ *Migration library, agent communication library, AWT (GUI) library*: These libraries are the default mobile agent toolkit's libraries.

BDI agents can be programmed using plan scripts. A definition of the simple plan script programming language using EBNF is given in Figure 5.5. In the plan language, each line of statement is represented by an action component. An action component could be any of the following:

1. A condition statement (IF-THEN)
2. A loop statement (WHILE)
3. A function statement (FUNCTION)

The function statement can be any of the following:

1. *An agent function*. This represents the agent's built-in functions; for example: move, askagentbelief, and so on.
2. *A Web service function*. A function to invoke a Web service.
3. *A user-defined function*. A function that represents a call to pre-compiled Java code. This function is useful for the agent programmer who wants to extend the CALMA agent's built-in functions.

The plan is defined in the <plan></plan> element. An agent designer can define context conditions where the plan is valid for execution. The context conditions are given within a <context-condition></context-con-

```
Plan                        ::== Goal Name Description
                                 ContextCondition+
                                 ContextFailure+
                                 Action+
Goal                        ::== STRING
Name                        ::== STRING
Description                 ::== STRING
Action                      ::== Function
                               | "WHILE" expression "DO" Action+
                                 "ENDWHILE"
                               | "IF" expression THEN Action+
                                 "ENDIF"
Function                    ::== <INVOKE> Agent_Function
                               | <INVOKE> Web_Service_Function
                               | <INVOKE> User_Defined Function
Agent_Function              ::== Method_name  Method_parameters
Web_Service_Function        ::== url   Method_name
                                 method_parameters
User_Defined_Function       ::== Class_name  Method_name
                                 Method_parameters
Class_name                  ::== STRING
Method_name                 ::== STRING
Method_parameters           ::== variable*
ContextCondition            ::== "CONTEXT" belief_fact
                                 "ENVIRONMENT" env
                                 ("FAILUREID" "=" failid)*
env                         ::== "arm" | "x86"
ContextFailure              ::== "CONTEXTFAILURE" "NAME" = failid
                                 Action+
url                         ::== "HTTP:" STRING
method_name                 ::== STRING
method_parameters           ::== expression*
belief_fact                 ::== variable operator <value>
expression                  ::== value | variable
value                       ::== STRING | INTEGER
variable_array              ::== variable+
variable                    ::== VARIABLE
Subgoal_name                ::== STRING
Operator                    ::== "lt" | "eq" | "gt" | "le" | "ge"
```

Figure 5.5 Plan language for Web service calls.

dition> section. In the case of failures happening within the particular context, the agent designer can define a set of actions in the <context-failure></context-failure> section. The <action></action> section in a plan specifies the agent's actions.

5.4.2 Server Component

The server component includes CalmaAgentManager, MatchMakerAgent, and PlanManager. To enable a task agent to request a service or a plan, the MatchMakerAgent Web service and the PlanManager Web service are provided.

The CalmaAgentManager services incoming requests from the mobile device component. It provides the list of available services, contacts the MatchMakerAgent to look for a task agent that can provide the requested services, and contacts the task agent for the requested services. The CalmaAgentManager will create and send a messenger task agent to notify the mobile user if no suitable task agent is available in the server agency.

The MatchMakerAgent enables the task agents to advertise their services. It maintains the list of advertised services and the associated task agent information. Task agents may deregister their services from the MatchMakerAgent.

Task agent plans are registered in the Plan Repository. Based on the service name, the PlanManager retrieves the plan from its repository. New plans can be added to the repository dynamically and the PlanManager can feed these to respective task agents at runtime.

5.4.3 Mobile Device Component

The mobile device component contains the MDAgentManager. It provides an interface for the user to request services and to manage task agents on the mobile device. The user might remove the task agents currently running on the device. Based on the task agent creation parameter, the agent is either removed from the device or is returned to its home agency.

When the user requests a service, the mobile device component first determines if a task agent for that request is available locally before contacting the server component. If the task agent is not present locally, the server component is contacted to find a match for the service as advertised via the MatchMakerAgent. The server component then contacts the CALMA task agent to perform the requested service. This task agent moves to the mobile device, reads the device's context condition, selects a plan, and starts executing the actions defined in the selected plan.

5.5 PROTOTYPE IMPLEMENTATION AND EVALUATION

In this section, the implementation of the CALMA infrastructure is illustrated, using a scenario. An evaluation of the infrastructure is presented in terms of lightweight and context-aware behavior that the implementation should support. Runtime execution performance is also discussed. The CALMA infrastructure is implemented using the Grasshopper Agent Platform, Apache Tomcat Web Server and Apache Axis Web Services.

The CALMA BDI Engine including the plan language and parsers have been implemented as add-ons to the Grasshopper and Aglets Mobile Agent

Toolkits. Enabling Web services through the PDA requires a SOAP protocol handler. The kSOAP package from Enhydra software (http://ksoap.enhydra.org) is used. The kSOAP package provides an interface between J2ME (Java 2 Micro Edition) with MIDP to access Web services that run on HTTP. On the mobile device implementation, Java code is executed using the Jeode Embedded Virtual Machine (EVM) which is based on Sun's Personal Java Specification. The agent's access to Web services is explicitly specified using the plan language.

5.5.1 Illustrative Scenarios

5.5.1.1 Booking Movie Tickets

Figure 5.6 presents a screenshot of the Grasshopper BDI agent that executes a plan containing an action to call a simple Web service from the mobile device. The Web service is hosted in the Internet Information Server (IIS) Web server that runs the .NET framework.

A simple scenario using CALMA agents to book movie tickets is shown in Figure 5.7. In brief, the booking-movie-ticket scenario works as follows. A user wants to book a movie ticket from a handheld device. After submitting a Book Movie Ticket task, a Local Agent will automatically ask for user preferences for the movie ticket. The preferences include type of movie, time, and location. Once the user fills in the preferences, the Local Agent will break down its task and create new subtasks. Then, the Local Agent will ask for agents that have the capability of pursuing those subtasks. We also assume that for simplicity a cinema has two kinds of servers. The first server hosts the CALMA system and the second server hosts Web services. In the handheld device, a stationary CALMA BDI agent is the Local Agent responsible for handling the user input. When the user selects the Book Movie Ticket task, the CALMA Stationary Agent creates a CALMA Mobile Agent. The CALMA Mobile Agent is given two plans. The first plan is to move to the first server and ask the CALMA Local Agent on the cinema server to book the ticket. The other plan is to book a movie ticket by invoking a Web service on the second server, without using mobile agents. The second plan can be used when the first plan fails or if the cinema server does not support mobile agents.

5.5.1.2 Finding an IDD Calling Card

The user requests the FindIDDCallingCardAgent to find the cheapest IDD calling card provider. The user then gets the desired phone card from the shop without the trouble of searching the phone card list for the cheapest provider.

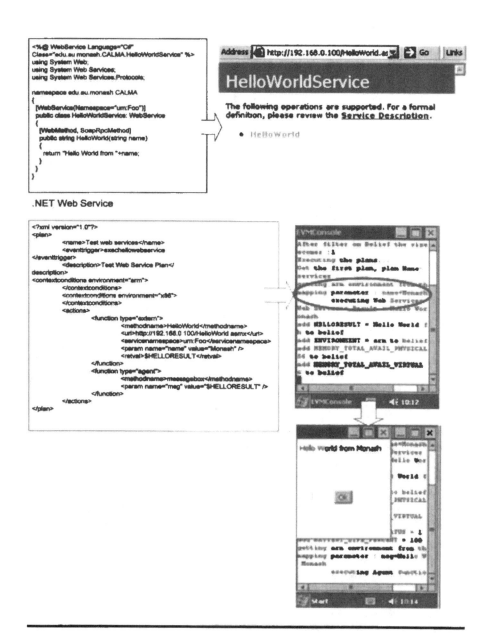

Figure 5.6 Executing an example Web service by a CALMA BDI agent.

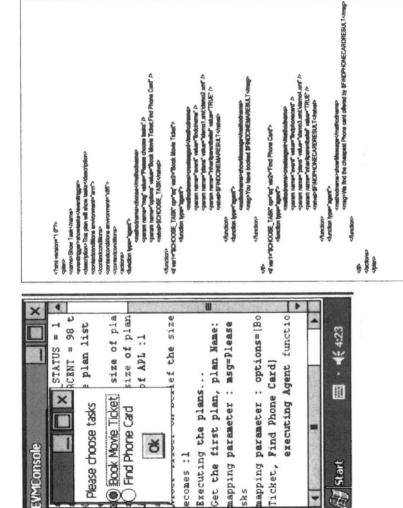

Figure 5.7 Movie booking CALMA agent.

Along with other task agents, the FindIDDCallingCardAgent is created via Grasshopper at the server side. The agent executes the actions in its start-up plan that includes invoking the MatchMaker Web service to advertise its service, and invoking the PlanManager Web service to load its plans.

In the FindIDDCallingCard scenario, the user selects the FindIDDCallingCard service from the list of advertised services as shown in Figure 5.8(a). As the CalmaAgentManager receives the request from the MDAgentManager, it informs the MatchMakerAgent to match the requested service with a task agent. The CalmaAgentManager then informs the FindIDDCallingCardAgent about the mobile user request. The FindIDDCallingCardAgent starts its execution by finding an applicable plan from its plan library to achieve its new goal. It first moves to the mobile device, and then reads and updates the device context information (updating its beliefs in the process) before it continues further actions.

When the FindIDDCallingCardAgent's intention is unable to continue based on its updated beliefs, the agent executes its default plan to request a new plan as shown in Figure 5.8(b). The default plan involves an action to invoke the PlanManager Web service. After retrieving the applicable plan from the PlanManager, it then starts executing the plan. In Figure 5.8(c), the mobile user submits the country name to the agent to initiate its search for the IDD calling card provider's name and call rate. The agent executes the actions in its plan and displays the results on the device.

5.5.2 Enabling Lightweight Behavior

As shown in Table 5.2, the size of the full BDI agent (with 10 initial plans, different plans catering for different contingencies) is almost three times larger than the lightweight BDI agent (with one initial plan, further plans downloadable if and when required). In resource-constrained environments, the lightweight agent model requires less memory to run and can perform the same tasks as the full BDI agent.

Besides the lightweight plan library, CALMA enables lightweight applications with respect to the mobile device (i.e., applications initiated on the mobile device avoid or reduce use of mobile device resources) via the following ways:

1. Enabling the task agent to move around in the network to perform its tasks autonomously (off-loading computations this way)
2. Enabling task agents to use external resources via Web service invocations
3. Reducing the size of the mobile device component by using the Grasshopper codebase to specify the task agent class library's location remotely

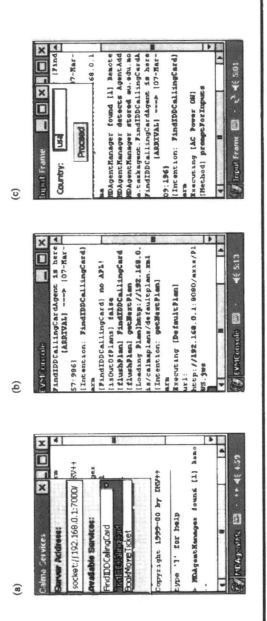

Figure 5.8 Finding an IDD calling card scenario.

Table 5.2   BDI Agent (with 10 Plans) and Lightweight BDI Agent Size Comparison

	Full BDI Agent	Lightweight BDI Agent
Size of agent	9050	9050
Total plan size	236560	23656
Size of BDI component	81544	81544
Total (in bytes)	327154	114250

Note: Size of plan depends on actions defined in agent's plans.

5.5.3   Enabling Context Awareness

In terms of the context awareness of the BDI agent, we have shown how FindIDDCallingCardAgent detects its execution context as it moves to a resource-constrained device. The agent reacts to its context situation and requests new plans from the server-side PlanManager. For the FindIDDCallingCardAgent to achieve its goals, it constantly reads and understands the context information while executing its current intention. It requests new plans from the server if its current execution is restricted in the new context.

The FindIDDCallingCardAgent also captures its migration and Web services invocation status. We enable the task agent to retry its migration and Web services invocation. The number of retries is configurable, based on the nature of the task agent. One task agent may be persistent in returning its result to the device user, whereas the other may quit after a number of attempts.

5.5.4   Performance Evaluation

In Figure 5.9, the performance of the lightweight FindIDDCallingCardAgent and the full FindIDDCallingCardAgent in the CALMA infrastructure are compared. The result is tested on the HP iPAC H4150 PDA. Figure 5.9(a) presents the results for experiment 1 (Exp1) in which the lightweight FindIDDCallingCardAgent contains the applicable plan when it migrates to the mobile device. Figure 5.9(b) presents the results for experiment 2 (Exp2) in which the lightweight FindIDDCallingCardAgent migrates to the mobile device without any applicable plan. The result presented is only indicative and would depend on the actual devices used, network bandwidth, and server-side processing power.

In Exp1, the lightweight FindIDDCallingCardAgent outperforms the full FindIDDCallingCardAgent in plan searching or loading. The full FindIDDCallingCardAgent uses a longer time to search through its plan library,

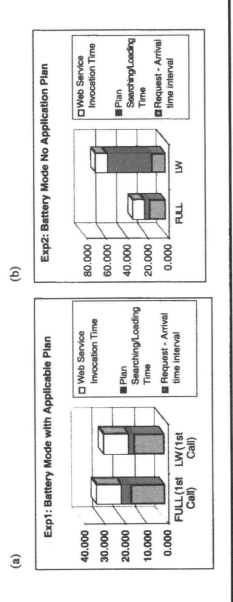

Figure 5.9 Execution performance results.

compared to the lightweight FindIDDCallingCardAgent. On the other hand, when the lightweight FindIDDCallingCardAgent does not have any applicable plan, as shown in Exp2, it requires a longer time to retrieve plans by invoking the plan request Web service.

Although the lightweight FindIDDCallingCardAgent requires more time to obtain an applicable plan compared to the full FindIDDCallingCardAgent in Exp2 (because it has to retrieve it from the server side), it is unlikely for even the full agent's plan library to contain plans for all possible execution contexts in the mobile environment. It is difficult to predict and specify plans to cater for all the possible context conditions. The (so-called) full FindIDDCallingCardAgent can still fail when it hits any unknown contexts and might still need to request other plans from the server side. Moreover, in the resource-constrained environment, the FindIDDCallingCardAgent is wasting device resources with its plan library that consists of a large number of nonapplicable plans. On the other hand, the lightweight agent not only is capable of performing the same task with a minimum requirement of memory, but it also enables dynamic updates of agents' plans. New sets of plans can be added to the Plan Repository dynamically to support executing task agents. In summary, whether agents travel with many or few plans, they might still need to retrieve new plans, and it would be useful to have an infrastructure to support such functionality.

5.6 SUMMARY

This chapter presented the CALMA framework to enable mobile agents operating in ubiquitous environments to be aware of changing contextual conditions and perform appropriate tasks — migratory or otherwise. The CALMA framework limits computation overheads to facilitate operation in resource-constrained devices by using a unique Web services model to off-load computations and by having a support infrastructure to facilitate obtaining plans "on demand" dynamically. Finally, the CALMA framework can be added to existing mobile agent toolkits as demonstrated via the prototype package implementation.

We have also seen other work on context-aware agents and Web service integration. CALMA is illustrative but does not exhaustively capture all possibilities with the concept of a context-aware intelligent agent. The key components are exemplified in CALMA and each component can be implemented using different technologies. For example, the BDI model is used, and other models can be employed such as rule-based reasoning or traditional AI planning, with their own merits and drawbacks. Also, only limited context information was demonstrated with CALMA, but other forms of context can be employed, even about the physical environment. The agent notion embodies and encapsulates a context-aware software entity, either incorporating the sensing, thinking, and acting subsystems, or utilizing such subsystems external to the agent.

ACKNOWLEDGMENTS

This chapter contains portions from (1) Sumartono, A., Loke, S.W., Krishnaswamy, S., and Chuah, S.H., Mobile BDI agent-driven interactions with Web services, *Proceedings of the International Conference on Intelligent Agents, Web Technologies and Internet Commerce 2004 (IAWTIC'2004),* Gold Coast, Australia, 2004, © *2004 IAWTIC2004, and (2) CALMA: context-aware lightweight mobile BDI agents for ubiquitous computing by* Chuah, S.H., Loke, S.W., Krishnaswamy, S., and Sumartono, A., *which appeared in the Proceedings of the Workshop on Agents for Ubiquitous Computing (available online at* http://www.ift.ulaval.ca/%7Emellouli/ubiagents04/acceptedPapers/11.pdf*).*

REFERENCES

Collier, R.W. and O'Hare, G.M.P., Agent factory: a revised agent prototyping environment, *Proceedings of the 10th Irish Conference on Artificial Intelligence and Cognitive Science (AICS)*, Cork, Ireland, 1999.

Dickinson, I. and Wooldridge, M., Towards practical reasoning agents for the semantic Web, *Proceedings of the 2nd International Joint Conference on Autonomous Agents and Multiagent Systems*, 2003, pp. 827–834.

Kotz, D., and Gray, R.S., Mobile agents and the future of the Internet, *Operating Systems Review* 33(3), 7–13, 1999.

Lowen, T.D., O'Hare, G.M.P., and O'Hare, P.T., Mobile agents point the way: context sensitive service delivery through mobile lightweight agents, *Proceedings of the 1st International Joint Conference on Autonomous Agents and Multiagent Systems*, Bologna, Italy, 2002, pp. 664–665.

Mihailescu, P. and Kendall, E.A., MAE: a mobile agent platform for building wireless M-commerce applications, *Proceedings of the 8th ECOOP Workshop on Mobile Object Systems: Agent Applications and New Frontiers*, Spain, June 2002.

Rahwan, T., Rahwan, I., and Ashri, R., Towards a mobile intelligent assistant: AgentSpeak(L) agents on mobile devices, *Proceedings of the 5th International Bi-Conference Workshop on Agent Oriented Information Systems (AOIS)*, Melbourne, Australia, 2003.

Sadeh, N.M., Chan, E., and Van, L., MyCampus: an agent-based environment for context-aware mobile services, *Proceedings of the Workshop on Ubiquitous Agents on Embedded, Wearable, and Mobile Devices at the International Conference on Autonomous Agents and Multiagent Systems*, Bologna, Italy, 2002.

Sumartono, A., Loke, S.W., Krishnaswamy, S., and Chuah, S.H., Mobile BDI agent-driven interactions with Web services, *Proceedings of the International Conference on Intelligent Agents, Web Technologies and Internet Commerce 2004 (IAWTIC'2004)*, Gold Coast, Australia, 2004.

Wooldridge, M., *An Introduction to Multiagent Systems*, John Wiley & Sons, New York, 2002.

6

CONTEXT-AWARE ADDRESSING AND COMMUNICATION FOR PEOPLE, THINGS, AND SOFTWARE AGENTS

In this chapter, we consider the use of context for enhancing communication among entities, including people, things, and software agents. Such use of context is often labeled *context-aware communication*, an area of practical importance. The general architecture is applicable; a subsystem gathers context information about entities, and then such information is used to enhance or regulate communication-related actions. As in systems seen earlier, the sensing and thinking subsystems can be decoupled and perhaps built separately from the subsystem that performs or supports communicative actions. However, we will explore examples of integrated systems in the sections that follow.

We have already seen a context-aware device for communication (i.e., the CAMP-UP system) and show how a device can be adapted via context information. In the case of CAMP-UP, the device is a communication device though the idea is applicable to other devices, including the coffee maker. This chapter explores further the use of context for people-to-people and people-to-object (devices, appliances, and everyday objects with embedded computers) communication, as well as communication between people and software agents, and communication between software agents.

127

6.1 CONTEXT-AWARE COMMUNICATION FOR PEOPLE

Information about an individual's location, activity, surrounding environment, and level of busy-ness can be used by communication technologies to provide more intelligent and adaptive support for interactions. The telephone with its amazing technology has helped communication since its invention in the 1870s, but it could ring at most inappropriate and inconvenient times and be a source of annoyance. Often, the callee does not have a choice and is forced to hear the ringing by the caller, even if the callee does not answer the phone. Now, in the 21st century, Schulzrinne (2005) describes one of the goals of context-aware communication as how to make the phone not ring except in an appropriate manner; that is, the phone should not ring in some callee situations but should ring in other situations.

Schilit (2002) noted several problems of today's ubiquitous communication technology, such as irrelevant communications, annoying disruptions, caller unaware that the callee is available for interaction, interaction overload (too much "junk" information or calls at the wrong time), and device overload (with the proliferation of device types), and defines *context-aware communication* as applying "knowledge of people's context (and activities) to reduce person-to-person communication barriers."

6.1.1 Application Types

The use of context for five application types has been noted in Schilit et al. (2002):

- Routing: Messages or voice calls can be routed to the appropriate destination. For example, if the location of the callee can be determined, the phone in the right room can ring (and in a distinctive tune) for the person. Also, call forwarding can be controlled based on other context attributes such as who they are with, where they are, what they are doing, and what the time of day it is. The idea of routing messages or calls to the appropriate device depending on user context is developed in the Ubiquitous Message Delivery (UMD) (Spreitzer and Theimer, 1994) and Mercury systems (Ranganathan and Lei, 2003). The UMD system will send the message to an appropriate terminal display or wait for a suitable situation or terminal to be available. The Mercury system can provide the following scenarios:
    - Route an incoming call to a one-way device if a two-way device is not available. For example, the call recipient who is away from the office and not running an instant messaging (IM) client can be paged.

- Automatically migrate a call to a different device because of a change in context during the communication session. For example, a person talking on a cell phone on the way to work can continue the conversation on a desktop IM client after arriving at the office.
- Exploit call priorities: For example, a highly anticipated call can be allowed to come through, causing the current call to be placed on hold.

- Addressing: Entities can be addressed based on their context instead of their unique identifier or names, as is done traditionally. This can be convenient when the identifiers are not known in advance or when this is simply easier. Two examples in Schilit et al. (2002) are a context-aware mailing list system, in which one can specify that a message be sent to "all people at a given location" or members of a research group currently in a building, and the PARCTAB virtual whiteboard, in which a group of people is virtually delineated by common location and membership in a project, so that all individuals in such a group can effectively view similar items.

- Messaging: By knowing the situation of the receiver or addressee, the right message can be delivered at the right time and in the right form, be it a reminder message or an incoming e-mail. For example, the MIT project Hanging Messages[1] (Chang and Maes, accessed in 2001) enables messages to be received at the right location and time. This application type is different from routing in that the message might have been routed to the right room (say, in which the receiver is currently located), but the message may still need to be delivered in an appropriate way according to the receiver's current situation in the room (e.g., the message must be delivered in a nonintrusive manner if the user is in the room and in a meeting). The Context-Aware Messaging Service (CAMS) system (Nakanishi et al., 2000) takes into account the addressee's communication context, including the schedule, location, and available media support, and selects the most suitable phone number or e-mail address for redirecting incoming messages. CAMS integrates a computer telephony integration (CTI) server, a positioning system based on Personal Handyphone System (PHS) and a scheduler (which indicates the user's activity at a given point in time, assuming that the schedule is indeed followed). The user carries a PHS handset and a PDA. A user can register rules about what to do with messages. For example, rules can be specified for the

---

[1] http://alumni.media.mit.edu/~elchang/HM/.

system to redirect calls to a voice messaging system when the user is lecturing at a university or having a meeting and to redirect messages to the right e-mail addresses. The user can edit his or her schedule via a form. The comMotion system (probably one of the earliest such context-aware communication systems) (Marmasse, 1999) aims to deliver reminders, to-do lists, messages, and other information in the most timely and relevant context. The message engine can also deliver news, weather reports, and traffic reports from Web sources to the user. An example of timely information delivery would be the user requesting a list of movies showing at a local cinema when leaving from work on a Friday evening. In the system, user position is obtained via a GPS system, and locations (GPS data) are labeled by the user as "home," "work," "grocery store," and so on.

■ Providing awareness of callee or receiver: By understanding the situation of an individual, one can determine if the individual is available to talk. Such presence awareness is similar to the idea of status in Instant Messenger systems. With the Live Contacts system (Henri ter Hofte et al., 2004) running on pocket PC phones, one could see presence information of listed contacts, including the current Messenger status (e.g., busy, online, away, on-the-phone, etc.), current calendar information (e.g., work and nonwork times, appointments, etc.), and last-known location information (e.g., at work, at home, on the move, or unknown). On selecting a contact on the phone's interface display, one can initiate contact (via telephone call, SMS, IM, or e-mail) or ask to be reminded to make contact at a nominated future time. The system is based on a client–server architecture with the devices querying Live Contact servers for updates on contacts. Collaborative work applications may also utilize presence awareness to optimize work interactions. Also, VoIP spam or instant messaging (IM) spam can be prevented using such presence technology (Schulzrinne, 2005). Related work on using sensors to gauge human interruptibility (Fogarty et al., 2005) is also useful in this application type.

■ Screening: Filtering or redirecting calls based on the situation of the receiver relates to the previous application type. In this case, it is not only that the system informs the caller of the callee's situation but also takes action automatically on perceiving that the callee is unavailable. We have this idea in use for context-aware mobile phones, where if the user (with his or her phone) is in a meeting, then the phone can be put in a certain mode, virtually filtering the call.

Note that the aforementioned application types may be combined in a specific application. For example, a system could route messages using context as well as filter appropriate messages. The Connector system (Danninger et al., 2005) maintains awareness of the users' activities and social relationships to help mediate proper connections at the right time. Presence, messaging, and filtering are supported; the system provides users with contextual cues about availability of callees and uses context to adapt the behavior of devices to avoid inappropriate interruptions. The idea of Connector is to facilitate appropriate and opportunistic connections for both parties. Context information used includes address book with contact information, user preferences and settings, and three other types of information (whether the user is in the smart room, whether a meeting is taking place in the room if the user is in there, and whether the user is outside the smart room and in a vehicle). Face recognition technology, vision-based techniques, and acoustic signals captured by a user-carried microphone are used to identify the user's context and current environment. Four interaction levels are supported: "available for talk," "available for instant message," "available for e-mail," and "available for voice mail." A distinctive feature of this system is the use of proactive software agents, where, if a contactee becomes available, the agents automatically find out if there are pending contacts to make connections.

Moreover, as in Schilit et al. (2002), different systems might have different levels of autonomy in acquiring context — the user might specify this, or the system automatically detects this via sensors — and in performing the action, given that the context is known; action is taken automatically or manually. These two dimensions provide a means to classify context-aware communication systems.

6.1.2 Call Services

Although communication in the aforementioned systems can mean IM, telephone, e-mail, or other mediums, there has been work specifically on using context for handling phone calls from the perspective of call services. A technique for implementing context-communication services for IP telephony is presented by Gortz et al. (2004). The system enhances the Session Initiation Protocol (SIP),[2] a signaling protocol for session and call control with context-awareness capabilities. Control of services is specified in scripts written in the XML-based Call Processing Language (CPL). Calls can be diverted or passed through to a colleague, depending on the context of the callee according to conditions specified, using special tags,

---

[2] http://www.cs.columbia.edu/sip/.

in the scripts. Next-generation mobile phones that use SIP can utilize such functionality, or a third-party call control proxy server can be used. Context information employed include indoor location sensing technologies (including Bluetooth, infrared/radio frequency badges) and the iCalendar-compliant applications, and are acquired via a context server consulted when a phone call arrives; an incoming call executes the CPL script, which, in the course of processing, calls the context server. A similar system using a SIP proxy and a comprehensive context manager is presented by McFadden et al. (2005). Context employed include not only location but also the status of the user, communication mode (e.g., voice, text, or video) available to the user, and device battery power. Context information is represented in the Context Modeling Language (CML). Such context-aware communication has also been implemented for VoIP (Kanter and Gustafsson, 1999). Context-aware communication to incorporate context into service provisioning and service adaptation is also the aim of the European Autonomic Communication project.[3]

6.1.3 More Applications

Context-aware communication systems can go further and perhaps provide an on-the-fly translation facility implemented on a device carried by a user if the language abilities of the speakers and listeners can be automatically ascertained via some mobile profile exchange mechanism. Some communication systems go further and aim at providing not only a translation facility but also context-dependent background knowledge to aid conversations, as in the Telme wearable computer system (Sumi and Nishida, 2001). Context is useful in a translation system in selecting the appropriate target language and in providing background knowledge suitable for the conversation at hand. In Telme, a novice and an expert can have a conversation, with the novice being aided by knowledge from the system.

There has also been early work on using such context-aware communication profiling to aid interaction with severely disabled users (Davis et al., 2003). The system stores information from conversations, such as topics last discussed and mood last generated. Such historical information and profiles of people provide context to guide future conversations. Automatic translation of speech into sign language or text as displayed on a device can also be of use in talking to deaf persons.[4]

---

[3] http://www.autonomic-communication.org/.

[4] See http://www.aaai.org/AITopics/html/machtr.html for links to projects that translate English into sign language.

6.1.4 Summary

In summary, there is tremendous opportunity in using context to enhance people-to-people communication. We have reviewed some work in the area, but many prototypes have been developed embodying particular ideas. However, there has not been extensive commercial exploitation of the ideas. Moreover, acquiring suitable context information is still very much a research topic, identifying not only what context information can best be used (e.g., in terms of cost and ease of deployment) to infer a user's situation but also how to best acquire such information, integrating appropriate sensor information and information resources (e.g., calendar and electronic diary). There is also the issue of reliability and accuracy in inferring the user's situation and what can be done if the system guesses incorrectly the situation of the user.

6.2 CONTEXT-AWARE ADDRESSING AND COMMANDING FOR OBJECTS

We now turn to artifacts and how context can be employed for people to naturally address or interact with such smart artifacts.

6.2.1 Application Types

Similar to the spirit of addressing in people-to-people communication, context can be used to address artifacts and to group artifacts together. For example, one could issue commands such as "turn off all appliances in the living room" and "turn down the heater in my son's room." In the former command, common location context is used to group devices together as well as to identify which devices to turn off. The latter command needs a knowledge base of context information to be understood. If each device knows of its own lifetime, one can ask all the devices in a living room, "which of you is more than three years old?"

Commands can similarly be issued to devices to be performed (or delivered to the device) only in an appropriate context of the device. For example, we can issue a command to a new advanced television to "turn off if no one is in the room after the news has been recorded," and the command "turn off" will be performed in the right context.

Hence, we note at least two application types: *addressing* and *context-aware commands*. We can have commands that utilize both these aspects, such as "when the 8 p.m. news is finished and if no one is watching, turn off the television next to the chair John is sitting in." In this command, the television is addressed by its context ("next to the chair John is sitting in") assuming that there is more than one television set, so that context is used to distinguish one from another, and the command "turn off" is only performed in the right context ("no one is watching" and "after the 8 p.m. news"). Context ensures that commands are delivered in the right way and at the right time to appliances. Proper interpretation of the commands is needed to determine which context attributes are being referred to in the command, and, once the particular attributes are determined, the appropriate sensing and reasoning subsystems can be employed. An infrastructure that supplies context information such as those mentioned in Chapter 2 can be used in these application types.

6.2.2   A View from Situation Semantics

A topic of linguistic theory provides some clarity about the nature of such commands. Commands are typically issued by humans in a particular situation, in which the command is given in the form of speech, gesture, via some device such as a keyboard or mouse, or some combination in a multimodal fashion. In the relational theory of meaning, or the situation semantics (Barwise and Perry, 1983) approach to giving meaning to utterance, the situation of utterance or the discourse situation is made explicit. An utterance W has meaning as a relation denoted $[W]$ between the discourse situation d and a mapping c, and another situation e (roughly representing the content of the W in context d) represented by a tuple of the form $d, c[W]e$.

In other words, human utterances (whether a command, assertion, or some other form of speech) have always been within some physical-world situation. And humans are able to perceive such situations (most of the time) and use them to understand utterances in their proper context. In the understanding of such utterances (or commands) by machines, the challenge is in the machines' perceiving the context surrounding the utterances. Perceiving context by machines is exactly the aim of context-aware computing.

Such a view presents a strategy for interpreting, in context, commands issued to devices, appliances, or everyday objects with embedded processors. Suppose W is a command uttered by a person and an infrastructure has sensors to determine the situation of the utterance, including, say, which room the user is in, what devices surround the user, and which device the user is likely to be issuing the command to (via the orientation of the user's body, what the user is looking at, pointing at, or touching). The mapping c maps particular nouns and verbs in the user's command to what might be referred to in the user's context. For example, a person issues the command "turn off" while looking at a particular lamp. With appropriate sensing mechanisms, a system can use context to translate the command into an appropriate operation on the lamp. We are assuming that the lamp exposes an API (e.g., hosting Web services in an embedded server) with which commands can be issued from a system to the lamp. The sensing and thinking subsystem figures out that the person is looking at the lamp, and the acting subsystem invokes a service call to turn the lamp off.

There is also a sense in which such commands are polymorphic, akin to how a method may be polymorphic in object-oriented languages, in that the method in an abstract class takes on the meaning of the method as defined in the concrete subclass. This also corresponds to what linguists call *efficiency of language*, in which the same phrase can be reused in different situations with similar meanings but different results depending on the situation at hand. For example, the command "switch on lights," when issued in different rooms, has similar meaning in the regard to what the command issuer intends, but probably does not mean exactly the same, as the command translates to a particular operation when issued in one room and to a different operation when issued in a different room (because each room has its own lights). It is the situation of utterance that distinguishes one use of a phrase from another use of the same phrase.

For illustration, let us consider simple phrase commands for objects of the form <operation> <device>. Examples of such simple commands are "turn on lights" and "open drapes." Such commands may be an utterance by a person "heard" by the system or input via a textual or other interface. The issue then is how a system would make sense of the command.

We identify four stages a system should support for processing these device commands:

1. Supply an interpretation: The commands must be understood by the system in some way and assigned semantics; one way to do this is to consider the situation (including the identity of the person issuing the command, the circumstances of this person, where the person is located, etc.) in which the command is issued. Consider the command "turn on light"; the system needs to figure out which light the user would have in mind (perhaps attempting to discover if such a device is in the user's current environment) and what operation on the light "turn on" describes.

2. Explore the effects of the command: It is useful to understand the effects of the command before they are executed; the idea is that the system can reason with these commands and assess their possible effects before actually executing them, for example, to guard against undesired behaviors or to determine the likelihood of success.

3. Execute the command: The system executes the commands based on its interpretation of the command and its judgment that the effects are "safe."

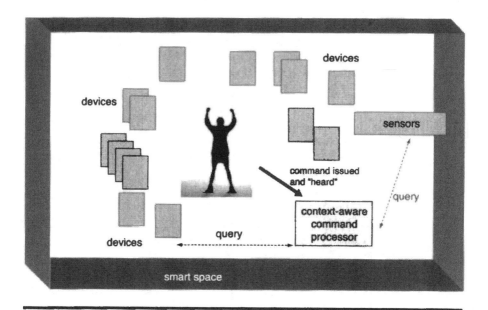

Figure 6.1 Illustration of a space in which there is a context-aware command processor.

4. Ascertain if the intended command succeeded: The system can check to see if the command produced the intended effects, perhaps by perceiving the situation after the command is completed (e.g., checking light sensors if the command is to turn the lights on), by obtaining status information from the device itself, or both ways.

Figure 6.1 depicts a space (simply termed a *smart space* in the sense that it can intercept and understand user commands) in which there is a processor for context-aware commands that attempts to execute commands issued by the speaker. The processor interprets the user's commands, translates it into an operation on a device, and invokes the operation on the device. It also queries sensors and devices to ascertain if a command has been successfully completed. In practice, the processor can be a server running in each smart room that continuously listens for user commands and has access to devices. We are assuming that the controllable or commandable devices have an API or a Web service-like interface on which operations can be invoked. Future devices might be controllable via an embedded Web server hosting Web services. Also, UPnP[5] provides such an interface to devices (e.g., the example of a toaster

---

[5] http://www.upnp.org/.

with a programmatic interface in Jeronimo and Weast (2003), and Sun's Jini provides a Java-based object model of devices.[6]

Although, to the author's knowledge, no such context-aware command system has been widely deployed, the context-aware speech system called CASIS (Leong et al., 2005) comes close, using the user's utterance situation to help in speech understanding. In CASIS, physical context acquired via sensors are used to disambiguate parts of speech in potentially ambiguous commands (e.g., "switch on that") to control devices within a room. The problem tackled in CASIS is formulated as computing the action (among a list of possible actions) with the highest probability, given the speech input and the current context of the speaker. An experiment was carried out with devices in a room, such as lights, projectors, projector screens, a video output switcher, a blind, and a DVD player, all controllable via speech commands. A total of 35 possible actions involving these devices, such as switching on the lights, lowering screens, switching on the projector, etc., were permitted. Context information used in the experiment was command history, device status, brightness in the room, speaker location, seat occupancy, sound level, and speech direction. The experiment showed how context can be used to resolve ambiguous commands with roughly 50 to 70 percent accuracy.

Using context for disambiguation may work in particular situations but is difficult to address in general. For example, imagine stepping into a room in which there is a table lamp, several ceiling lights, and a set of wall lights. Stepping into a completely dark room, the user could issue a command such as "switch on lights," which is ambiguous and could mean all the lights or only a particular light. Or, perhaps, in this case, even the user is unsure of what is intended, as long as there is some light. In such commands, perhaps a default behavior needs to be instrumented in the system, which can differ from one situation to another (or one room to another), or adjustment commands can be employed (e.g., "more," "less," "reduce").

Such commands are reminiscent of science fiction movies in which humans issue commands to devices or the computer to perform tasks, and the system is able to figure out which device is involved and what operation is being intended on the device. CASIS and works of others such as Loke et al. (2004) are initial attempts toward such systems.

6.2.3 Summary

In summary, the aim is to be able to command appliances, devices, and everyday artifacts naturally, and machine-understanding of context of

---

[6] http://www.jini.org/.

utterances can aid this endeavor not only in the interpretation of the content of the command but also in identifying which device the user is addressing the command to and what kind of operation is being intended.

6.3 CONTEXT-AWARE COMMUNICATION FOR SOFTWARE AGENTS

Intelligent software agents is an active area of research, and interagent communication with specialized languages based on speech act theory have enjoyed much attention (e.g., Labrou et al., 1999). There has been work on routing messages to mobile agents, which is certainly more challenging than in the case of stationary agents. This section focuses on addressing agents.

6.3.1 Addressing Agents via Context

Software agents are usually provided with a system-generated unique identifier for addressing. In the rest of this section, we consider the issue of using context to address software agents, in particular, mobile software agents, which adds the complexity of mobility and therefore the need to update changing context information about such agents. In principle, it is not difficult to imagine the use of the platform and mechanisms we consider here for addressing objects (devices and appliances) and people, but the illustration here involves messages based on an event notification system for software agents, first introduced by Loke et al. (2003).

Although we allow our agents to be mobile, not all agents require such a property, but agent mobility has been recognized as beneficial in a number of large-scale distributed applications, and a sizable number of mobile agent toolkits have been developed for mobile multiagent applications.

The ability to address or refer to software agents using a variety of methods can provide flexibility and abstraction for system developers, particularly when the system being built is dynamic and large, involving a huge number of agents, and in the case of applications in which the agents are intended to inhabit computers embedded in physical environments (e.g., agents roaming over hosts throughout an intelligent building as intended in the Hive project[7]).

Context can be used to send messages to a collection of agents based on their current context (even the physical context of the agent hosting computers), without knowing the precise identities of the agents. For example, as noted in Loke et al. (2003), we can send a message to all

---

[7] http://hive.sourceforge.net/hive-asama.html.

agents currently on hosts situated on the second floor; alternatively, we want to send a message to all agents launched by, and belonging to, a user, to all agents currently running on hosts with decreasing bandwidth, to all agents currently on hosts that provide a particular kind of E-service, to all agents that have not yet completed their tasks, or to the host nearest to a particular user. We can also ask agents for their identifier via a query such as "send me your name if you are on a host on the third floor." Such context-based addressing and messaging will be useful for distributed monitoring and management applications.

The mechanism in Loke et al. (2003) for context-based addressing of agents is an event-notification system called Elvin and relies on agents truthfully and proactively reporting, or notifying the event server of, their current context. The event server then forwards messages to the agents by matching the context as reported by the agents with the context specified in the message. Elvin uses the mechanism called *content-based addressing*, in which notifications (a set of attribute-value pairs) are forwarded to subscribers whose subscriptions (in the form of string-matching expressions on values of attributes) match the content of the notifications. For example, a message intended for agents running on the host nearest to a particular user will be sent to the agents reporting that they are on such a host. The actual context utilized and the reporting frequency will depend on the specific agent application. Agents can subscribe to the server for particular messages and listen for messages intended for a particular context. Or a user who wants to send messages to the agents can do so by sending a notification to the central server. Figure 6.2 illustrates the architecture of agents whose messaging is supported by the Elvin server.

The mechanism described earlier has been used to define namespaces in an ad hoc fashion over collections of agents. For example, one can create a namespace that relates to the geographic position of machines hosting agents. A building can be divided into logical areas such as floor1, floor2, floor3, floor4, and floor5, and each floor can be subdivided further into rooms, room1, room2, etc. One can then refer to "agents currently in floor4 OR floor5, who belong to Jack" AND send messages to such agents.

6.3.2 Applications

Such a mechanism can be explored for addressing robots and, in particular, swarms of robots, where it is impractical to individually address each one by name. Of course, the use of context-based addressing is merely an abstraction — each robot or software agent still requires a unique system-given identifier (e.g., an IP address for a computer running on the robot)

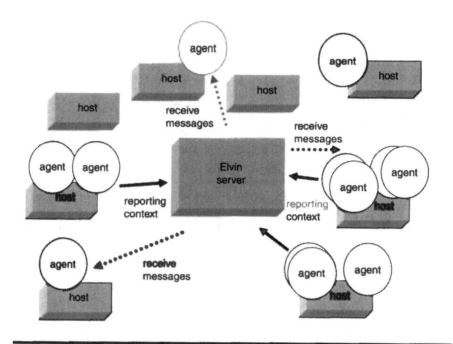

Figure 6.2 Elvin content-based addressing for agent context-based addressing.

for messaging purposes. For example, we can envision teams of robots searching for survivors in a tsunami-hit city, where one could refer to such robots by their current location, what objects they are currently in close proximity to, their current activity, current power levels, capabilities they have, or current findings. Wireless connections to an event-notification server will be useful with robots continually reporting their context via a large number of small messages. For efficiency, deltas or changes in context can be reported rather than repeating previous unchanged reports. Similarly, a mixed team of human workers and robots might be addressed via an ad hoc namespace formed using their current contexts. We also note that the event-based mechanism described here can also be adapted for people and objects.

6.4 SUMMARY AND CONCLUSION

In this chapter, we have presented different ways in which context can be employed to enhance communication between people, between people and artifacts, between people and agents, and between agents. These methods reflect the general idea that communication typically occurs within some context rather than in a vacuum, and this context can be captured and exploited.

REFERENCES

Barwise, J. and Perry, J., *Situations and Attitudes*, Cambridge, MA: MIT-Bradford, 1983.

Chang, E.L. and Maes, P., Hanging Messages: Using Context-Enhanced Messages for Just-In-Time Communication, available at http://alumni.media.mit.edu/~elchang/HM/docs/hm_brief.pdf.

Danninger, M., Flaherty, G., Bernardin, K., Ekenel, H.K., Kohler, T., Malkin, R., Stiefelhagen, R., and Waibel, A., The connector — facilitating context-aware communication, *Proceedings of the ICMI*, Trento, Italy, October 2005, ACM Press.

Davis, A.B., Moore, M.M., and Storey, V.C., Context-Aware Communication for Severely Disabled Users, CUU, Vancouver, British Columbia, Canada, November 2003, ACM Press.

Fogarty, J., Hudson, S.E., Atkeson, C.G., Avrahami, D., Forlizzi, J., Kiesler, S., Lee, J.C., and Yang, J., Predicting human interruptibility with sensors, *ACM Transactions on Computer-Human Interaction (TOCHI)* 12(1), 119–146, March 2005.

Gortz, M., Ackermann, R., and Steinmetz, R., Enhanced SIP communication services by context sharing, *Proceedings of the Euromicro 2004 Conference*, August 2004, pp. 272–279, available at http: phoenix.labri.fr. documentation/sip/Documentation. Papers. Programming_SIP. Paper_Publication_and_Draft/gortz.pdf.

Henri ter Hofte, G., Otte, R.A.A., Kruse, H.C.J., and Snijders, M., Context-aware communication with live contacts, *Proceedings of the Conference on CSCW*, Chicago, IL, November 2004, ACM Press.

Jeronimo, M. and Weast, J., *UPnP Design by Example: A Software Developer's Guide to Universal Plug and Play*, Intel Press, U.S.A., 2003.

Kanter, T. and Gustafsson, H., VoIP in context-aware communication spaces, *Proceedings of the 1st International Symposium on Handheld and Ubiquitous Computing (HUC)*, 1999, Lecture Notes in Computer Science 1707, Springer-Verlag, pp. 365–367.

Labrou, Y., Finin, T., and Peng, Y., Agent communication languages: the current landscape, *IEEE Intelligent Systems* 14(2), 45–52, 1999.

Leong, L.H., Kobayashi, S., Koshizuka, N., and Sakamura, K., CASIS: a context-aware speech interface system, *Proceedings of the 10th International Conference on Intelligent User Interfaces*, 2005, ACM Press, pp. 231–238.

Loke, S.W., Padovitz, A., and Zaslavsky, A., Context-based addressing: the concept and an implementation for large-scale mobile agent systems using publish-subscribe event notification, *Proceedings of the 4th IFIP International Conference on Distributed Applications and Interoperable Systems (DAIS 2003)*, Paris, Stefani, J.-B., Demeure, I., and Hagimont, D., Eds., 2003, Springer-Verlag, Lecture Notes in Computer Science 2893, pp. 274–284.

Loke, S.W., Stanski, P., and Syukur, E., Adding context-aware behaviour to almost anything: the case of context-aware device ecologies, *Proceedings of the Workshop on Context-Awareness at the 2nd International Conference on Mobile Systems, Applications and Services (MobiSys'04)*, Boston, MA, 2004.

Marmasse, N., comMotion: a context-aware communication system, *Proceedings of CHI 1999*, ACM Press.

McFadden, T., Henricksen, K., Indulska, J., and Mascaro, P., Applying a disciplined approach to the development of a context-aware communication application, *Proceedings of the International Conference on Pervasive Computing and Communications*, 2005, pp. 300–306.

Nakanishi, Y., Tsuji, T., Ohyama, M., and Hakozaki, K., Context aware messaging service: a dynamical messaging delivery using location information and schedule information, *Personal and Ubiquitous Computing* 4(4), 221–224, 2000, Springer-Verlag.

Ranganathan, A. and Lei, H., Context-aware communication, *IEEE Computer* 36(4), 90–92, 2003, IEEE Computer Society Press.

Schilit, B.N., Context-Aware Communication — Tutorial at Pervasive 2002, available at http://seattleweb.intel-research.net/people/schilit/Tutorial%20T3%20-%20Context-Aware%20Communication.ppt.

Schilit, B.N., Hilbert, D.M., and Trevor, J., Context-aware communication, *IEEE Wireless Communications*, October 2002, IEEE Press.

Schulzrinne, H., Making the phone not ring, presentation at the Internet2 Spring Meeting, 2005, available at http://pic.internet2.edu/20050504-smm-session/20050504-hgs-pic.pdf.

Spreitzer, M. and Theimer, M., Architectural considerations for scalable, secure, mobile computing with location information, *Proceedings of the 14th International Conference on Distributed Computing Systems*, 1994, IEEE Computer Society Press, pp. 29–38.

Sumi, K. and Nishida, T., Telme: a personalized, context-aware communication support system, *IEEE Intelligent Systems* 16(3), 2–8, 2001, IEEE Computer Society Press.

7

CONTEXT-AWARE
SENSOR NETWORKS

Sensor networks are increasingly important for numerous applications from smart kindergartens to intelligent traffic management as mentioned in Chapter 2 and Akyldiz et al. (2002), and in understanding regularity in everyday life behavior (Clarkson, 2002). There has been tremendous excitement about sensor networks in recent years with research into the different layers of technologies from networking protocols, hardware, and operating systems to programming models. Sensor networks are related to context-aware computing in at least two ways. First, sensors are used to acquire contextual information for entities. Second, sensors themselves can be made to be aware of their own context. This chapter introduces the notion of context-aware sensors whose behavior can be changed in response to their current context, thereby allowing power-conserving actions to be performed with the sensors depending on the situation the sensors are in.

We begin the chapter by introducing the concept of the context-aware sensor, making reference to related work. To illustrate how such a concept can be realized, we propose a software architecture for such sensors and then provide an application scenario to illustrate the architecture.

7.1 CONTEXT-AWARE SENSORS: THE CONCEPT

Each sensor comprises hardware with limited computational and networking capabilities. A *wireless sensor network* (WSN) is a combination of low-cost, low-power, multifunctional miniature sensor devices consisting of sensing, data processing, and communicating components, networked through wireless links. In a typical application, a large number of such sensor nodes are deployed over an area with wireless communication capabilities between neighboring nodes. Forming a mesh network, nodes can then relay sensed information back to base stations. Although low sensor costs, miniaturization, processing capabilities, and portability are enabling factors, energy constraints of sensors are a constant challenge because of generally slow progress in increasing battery capacity.

Activities of a sensor require energy (or use of battery power), whether for sending or receiving data or sensing. The idea with context-aware sensors is that if sensors could know more about their own context, then they could adapt their behavior and function only when needed and to the extent needed in the current circumstances, thereby becoming more prudent about how they should spend their energy.

The work of Elnahrawy and Nath (2004) proposed the novel concept of sensors, also labeled context-aware sensors, which can be aware of their history of sensed values, as well as of the history-sensed values of neighboring nodes. Aware of such readings, a sensor can construct correlations between its own sensed values and the sensed values of neighboring nodes as well as its own past values. Such correlations can then be used to predict future sensed values and to detect outliers and faulty sensors, and to fill in missing values. This awareness was not used to help reduce battery power.

The work by Cardell-Oliver et al. (2005) illustrates the context-aware sensor concept. In a sensor network's application to monitor soil moisture, the frequency of measurements taken varies according to the rainfall; during wet periods, measurements are taken more frequently, whereas during dry periods, measurements are taken less frequently. Rain sensing nodes (i.e., rain gauge) can detect a high rainfall event (say, exceeding 1 mm of rainfall), which is then used to determine how active the soil moisture probes should be. In this way, the soil moisture probes only need to sample more frequently when significant changes in soil moisture conditions are more likely. The advantage of this is the energy savings for the soil moisture probes. Given the large difference between energy consumption during the sleep (inactive) and active states of a sensor node (for a Berkeley mote sensor hardware, power draw can be 5 to 20 mA during the active period and 5 μA during sleep periods) and, overall, if less data needs to be transmitted, the energy savings can be many times greater than without such selective activity.

The same idea can be applied to other applications. For example, in the habitat-monitoring project at Great Duck Island (Mainwaring et al., 2002), sensor data patterns show that petrels are unlikely to enter their nests during the light phase of a 24-h cycle. This indicates that we would require less monitoring of activities in the nests during that period of time. Hence, adapting to this event, the monitoring sensors in the nests could reduce sampling rates during those periods. This approach is context-aware sensing because the sensor node is made to adapt its operations based on patterns or events (as perhaps detected by other sensors) that occur in the physical environment.

The proposed CASN (Context-Aware SensorNet) middleware (Huaifeng and Xingshe, 2005) takes seriously the idea of context-aware sensors that use available contexts and adapt their behaviors. They advocate the use of fuzzy rules to map detected contexts to actions. Although their work is still in the design stage, we note the use of context in CASN to reduce energy usage and to automate changes in sensor behaviors (using context and fuzzy rules).

The work by Khai et al. (2005) investigated a general framework for supporting such context-aware sensors with energy savings in mind, and the rest of this chapter describes this framework and an application scenario.

7.2 A FRAMEWORK FOR CONTEXT-AWARE SENSORS

As in many context-aware applications, context could be used to trigger information specific to users, providing users with a more informed decision space and aiding user decisions. One can apply the concept to sensor networks for reducing energy consumption and thereby prolonging battery life, as later demonstrated through an implementation, with results and applicability to real-life applications also included.

A simplified view of the stages in this process would involve analyzing sensor data streams for contextual information, mapping the discovered context into respective triggers, and later using the triggers to execute power management functions in concerned sensors where appropriate. The general idea underlying the context-aware sensor framework is (1) establishing patterns for controlling sensor activity (either programmed by developers or acquired by learning from sensor data) and (2) using the patterns as triggers of energy-saving operations on sensors. For instance, rules such as "in response to low habitat activity, put the sensors to sleep" or "in response to high rainfall, have the sensors highly active" might be programmed by application designers.

7.2.1 Sensor Roles

In the architecture, we assume that sensor nodes can assume different roles in a sensor network. For instance, we can assign certain motes (i.e., a sensor node that is a Berkeley mote) to only perform sensing operations and, at the same time, control other motes in their vicinity. Conversely, other sensor nodes might assume the role of sensing information only and not be able to give instructions. This is illustrated in Figure 7.1.

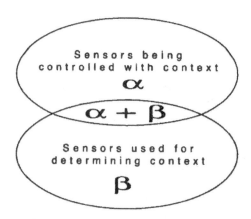

Figure 7.1 Sensor monitoring roles.

Such a simple role differentiation in a sensor network would enable us to power-save motes although the basic sensing operations are still preserved in different sensor patches or arranged sensor groups. Control instructions that are possible with sensor nodes range from minimum control by means of adjusting sensing properties to higher control by commanding motes to sleep for a fixed period of time, waking up only at subsequent clock cycles. The energy states of sensors that can be manipulated are further described in the following text.

7.2.2 Categorizing Energy Consumption

A finite-state machine (FSM) can be utilized to model the energy states of sensor nodes. Here, we review the main sensor power states and the types of commands that will yield a transition change. We also focus on expressing more generally the requirements that enable power savings in any sensor network to discover state transitions required to achieve maximum energy savings.

7.2.2.1 Input Alphabet

The current context when discovered in the system activates a triggering rule, enabling commands to be sent to prespecified sensor nodes. The input alphabet is represented by such commands, which can be understood by any type of sensor and differs for different sensor types. For mica2 sensor nodes, the input alphabet is formed by the set of the following possible commands:

1. Sleep <x seconds>.
2. Message size adjust <msg_type>.

Energy States

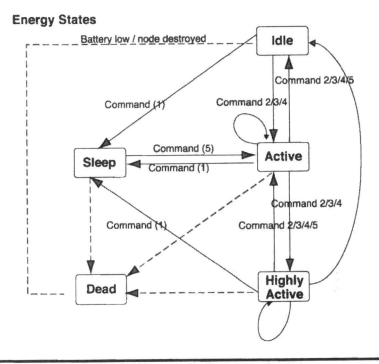

Figure 7.2   FSM for energy states.

3. Transmission frequency set <per y millisecs>.
4. Set radio strength <z hertz>.
5. Default operation.

Figure 7.2 shows the states and transitions, each transition labeled with one or more of the preceding five commands that cause the state transition.

7.2.2.2   Output Alphabet

The output of a state transition is the actual sensor operation on the sensor that results in a change of power state. Basically, in the FSM model, a command from the input alphabet causes an operation on the sensor, which, in turn, causes the change in power consumption states. Hence, in the model, we simply specify a direct correspondence between the input alphabet and the output alphabet. For instance, the sensor operation sleep <x seconds> could change a sensor's power state from active to sleep mode for a period of time. The output alphabet is therefore one low-level sensor operation or a collection of such operations that enables this change of state, or any operation that could cause a status change in the course of a sensor's current operation.

7.2.2.3 Energy States

A state in the FSM represents the current power state of a sensor, which is determined by the current command operated by the sensor. Here, the sensors represented include sensors being controlled with context, sensors used for determining context, or both. The state diagram is shown in Figure 7.2.

The five power states used here are as follows:

- Sleep: A power state with a fixed period of inactivity; for instance, motes sleep for maximum power savings. A mote in this power state will not respond to messages from other motes or the control application. It is assumed that for motes reaching this state, the triggered context is to have an effect only for a known period of time, and when the time limit has expired, the mote will exit this state and go back into normal operation mode.
- Idle: A sensor is in idle state when it is maintaining operation at a lower energy cost relative to its normal operating mode. For instance, it could have reduced communication frequency that reduces the number of messages exchanged between the server and the motes or, on the other hand, a reduced computation load when sensors aggregate data when possible and send packets of reduced sizes when it is known that that particular information is redundant. It is a mode whereby a sensor operates with a subset of minimum operating functionalities to stay alive.
- Highly_Active: A sensor can be highly active while monitoring a phenomenon, for example, to provide a higher sampling rate.
- Active: A normal mode, where motes are operating under normal conditions with no power savings.
- Dead: A dead state represents a mote that has failed in operation, for instance, when it is energy depleted or has died because of external factors such as being destroyed by a fire. This is not a power state that falls under the control of our system. It is an exit state for nodes that have failed in operation.

7.2.3 Architecture

The framework considers (1) a setting of groups of sensors that can provide context as well as send and receive commands (as described earlier), (2) several support services that provide communication between sensors and the system, and (3) components that provide the core functionalities for context discovery and use.

Programming of the sensor nodes is done in nesC, a variant of the C programming language for embedded sensor network devices under the

TinyOS operating system. To interface with the sensors, the Java 1.1 API is used for modules that provide data collection, analysis, and communication between the motes and the base station. This section describes the design and implementation of the individual components that constitute the system, and how the components interact. Figure 7.3 shows the architecture of the framework.

The major components of the framework are described as follows.

Communication server: The communication server acts as a mediator between the base station and the sensors; it receives raw data packets that arrive from the sensors and parses them into a format recognizable by the application. Similarly, any control messages from the application to the sensors are parsed at the communication server. In essence, the communication server provides a unified protocol of communication for applications (implemented in a high-level language), which need to send and receive commands from sensors implemented in a low-level language, TinyOS in this case.

Context locator service: To interpret the current context from the raw sensor readings obtained, the current framework employs preprogrammed if-else condition rules. The rules are used to decide whether certain context has been acquired. In the system, useful context information is mostly secondary context formed by conjunctions of different primary context information. For instance, if temperature readings fall below a certain threshold, the primary context is *cold*, whereas the useful secondary context could be *cold_weather* when sensors in a group detect the same primary context.

Context mining service: Ideally, however, context should be discovered autonomously in the system via data mining techniques. A current prototype of the framework (Chong et al., 2005) involves user-programmed rules, and so, context discovery is static and dependent on designed scenarios. Although not implemented in the current prototype, such a context mining service could mine for context from sampled sensor packets forwarded to it by the context locator service. The role of the context locator service would then be sampling sensor data and discovering context via the context mining service.

Context trigger engine: In the framework, the context trigger engine determines the sensor operations triggered, using a combination of given context and sensor profiles. This engine comprises three components: (1) context scheduler, (2) macro decoder, and (3) action interpreter. The core of the trigger engine lies in the use of action macros. Action macros are a command or a combination of actual commands sent to sensors.

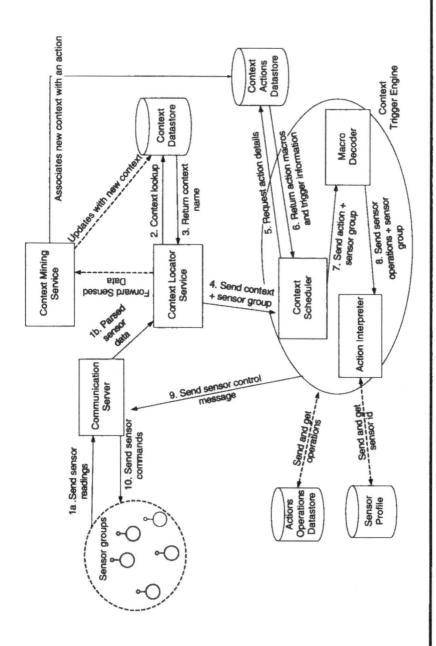

Figure 7.3 Architectural diagram showing the main components of the framework.

Context action datastore: As defined earlier, the relationship between context and its respective actions is either a one-to-one mapping or mappings with interdefined relations. In the prototype, these mappings are stored as strings. Sensor profiles and context names are stored in a file. Alternatively, an XML database could be used for storing the mappings information in a more precise format.

The following paragraphs describe each of the steps in Figure 7.3:

Steps 1a and 1b: Send sensor readings and parsed sensor data. At regular intervals, sensor nodes collect samples of data about the environment and send the packets to higher resource entities such as PDAs or laptops. The raw hexadecimal data packets are then parsed and interpreted by the system to construct equivalent decimal readings in a format that can be parsed by the application later.

Steps 2 and 3: Context lookup and return context name. Sensor data packets from each sensor group are then averaged when x number of samples have been collected, where x is a user-defined value based on the accuracy of readings within sensor groups. This yields the primary context for sensor groups. At each iteration of discovered primary context, the system updates the secondary context using the new primary context. The secondary context information is then passed on to the context trigger engine. If new context information is discovered, then this information would be passed on to storage, pending user response.

Step 4: Send context label and sensor group labels. The context trigger engine serves to match the current context to a possible if-then rule that can be used to trigger valid sensor operations. To match the context to an appropriate rule, other contextual information about the system can be used (e.g., the current time and the sensor groups that gathered the information).

Step 5 and 6: Request action details and return action macros and trigger information. Once a matching action macro has been found for some context, the system puts this macro on the queue to be scheduled for triggering. The queue operates based on a time-to-trigger parameter, which is a countdown timer to activate each individual action macro. When the timer reaches zero, that macro will be activated. This parameter is used to prioritize the macros activated to maximize the amount of energy that can be saved (for example, sleep mode may preempt less-active mode) and to accommodate time to establish that the context is valid. The time-to-trigger is user defined and decided based on the sensitivity of a context change.

Step 7: Send action and sensor group. Upon activation of a macro action in the system, the macro decoder component will decompose the macro into simpler instructions to be sent to the listening application running at a base station.

Steps 8, 9, and 10: Send sensor operations and sensor group, send sensor control message, and send sensor commands. At the application end, decomposed action messages would be mapped to actual sensor operations that sensors can understand. These sensor commands are then sent by the application to individual groups of sensors.

7.3 IMPLEMENTATION AND APPLICATION SCENARIO

The previous section described the framework for a context-aware sensor network. This section will highlight the inner workings of the implementation by Chong et al. (2005), using an application scenario that sets the context for the implementation. To implement the sensor network, Berkeley mica2 motes have been deployed as the sensor nodes. Each mote has the ability to process data, communicate with neighboring nodes through onboard radio, and contain various sensors. In the prototype application, however, only light and temperature sensors on the motes are used.

As an application scenario, let us envisage a pig farm with strategically deployed sensor nodes. Hence, all the sensor nodes have the ability to produce measurements for temperature, light, and sound in the locations where they are deployed. In this hypothetical scenario, these sensor nodes are programmed to collect data in packets, communicate this data to their nearest neighbor nodes, which would then upload the aggregated data to the base station or other motes in close proximity to the base station. Here, sensors that provide contextual information are sensors in the pigsties and sensors outside, although only sensors in the pigsties are being controlled. When the data arrives at the base station, it would be collected and analyzed for specific patterns.

Study of this data would yield a pattern, which can be checked with a context database to determine if it is a known context, and, if not, it is added as additional context into the database. As an example, if the sensors located in all pigsties detect no or minimum sound for long periods of time, this would be a new recognized pattern, and the identifying context could be "Pigs temporarily out of sties." The context data store would be automatically updated with context labels as new patterns are being discovered in the data, and at the same time, it would prompt the user for trigger actions for new contexts. Continuing from this example, when the system knows "Pigs temporarily out of sties," a context-triggered action would be to reduce sensing in concerned areas until the pigs come back (assuming that collected sensor readings will not be useful when there are no pigs). Another context-triggered action could be to reduce sensor message transmission.

In implementation, a separate data store could be utilized to provide the context to trigger actions, where triggered actions are an abstracted view of the low-level code running on the sensors. The scenario could be described in four stages:

1. In our scenario, upon a database lookup, the context "Pigs temporarily out of sties" maps to the action "reduce sensing."
2. Next, the decoder in place decodes the action macro "reduce sensing" into simpler action commands such as "reduce light sensing" and "reduce sound sensing."
3. In a one-to-one mapping, action labels map to sensor commands and an interpreter derives commands (by utilizing the code database and sensor profiles) to be sent to the communication server on the base station.
4. The base station receives commands and sends them to intended sensor addresses (i.e., to the sensors being controlled).

Upon receiving commands, sensors would adjust their behavior until the next context is sensed. In this scenario, a context such as "Pigs in sties" would reset sensors' sensing behavior to the intended default.

7.3.1 Experimental Investigations

The effectiveness of the framework for saving energy in wireless sensor networks has been evaluated via the prototype implementation of the components, and experiments have been performed using this prototype. All readings were recorded from actual runs of the application and graphs from simulations using PowerTOSSIM.[1] All experiments simulate the pig farm scenario described earlier, and, in all context-aware experiments, packets are written whenever the base station sends out sensor commands to individual motes. Specifically, the main objective of the experiments is to obtain an estimate of the total energy saved using our framework in terms of radio communication, data throughput, and voltage readings. In these experiments, the sensor network configuration assumed is as shown in Figure 7.4; a group of sensors play the role of providing context information for controlling another group of sensors (playing the role of controlled sensors). Details of the four experiments performed and their summarized results are given as follows:

---

[1] http://www.eecs.harvard.edu/~shnayder/ptossim/.

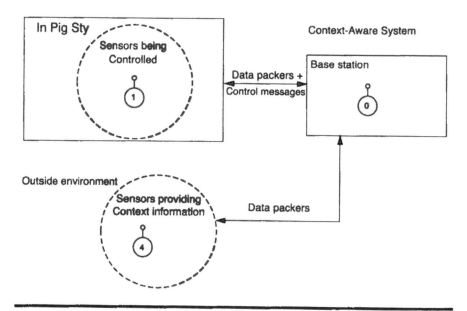

Figure 7.4 Experimental setup.

Experiment 1 — control experiment (as a benchmark for comparison): In this control experiment, the setup involves placing all motes under room temperature, where no context changes will take place over a period of 20 min. Motes 4 and 1 collect light, temperature, and battery voltage readings every 1000 ms and transmit every full packet to the base station. This experiment demonstrates the sensor network described in the aforementioned scenario in which the data acquired by the sensor is directly stored at the base station. Here, the system assumes a constant *context state* of Normal.

Experiment 2 — transmission rates experiment: Utilizing our context-aware framework, one action that is implemented is varying the transmission rates of packets in accordance with context changes, where the transmission interval t can take values of 1000 ms, 10000 ms, and 100000 ms for sensor nodes transmitting data packets per t ms. In total, 12 packets have been written out to sensor motes to change their transmission frequency, corresponding to changes among the three context states over the time period; the context states are (1) Pigs_Need_Out, (2) Normal, (3) Pigs_Need_In_Light. In comparison to the control experiment, the results show a reduction in total messages sent from motes of 33% (202/608 × 100), which equates to energy savings. The energy benefits are also indicated in the changes in voltage measurements when contrasting the two experiments.

Table 7.1 Table of Experimental Results

	Experiments			
Results Analysis	Control	Exp2	Exp3	Exp4
Runtime (in min)	20	20	7	20
Packets read	608	406	207	310
Packets written	0	12	11	10
Voltages (in mV) mote 1 start	2912	2912	2898	2892
mote 1 end	2892	2898	2892	2872
mote 4 start	2892	2892	2898	2687
Mote 4 end	2878	2885	2885	2653
Context states taken	0	3	2	2

Experiment 3 — message size experiment: As most energy is expended in radio communication, energy consumption would most significantly depend on the packets read and written. As such, the size of packets communicated in the network is a factor that can conserve energy. From a data management point of view, minimized packet sizes could be a product of data aggregation or data filtering activities in the network. This experiment aims to study the possible energy savings from reduced message sizes when context changes are experienced.

Table 7.1 shows the results from an experimental run of 7 min. Message size changes are not observed using PowerTOSSIM and hence not presented. During the experimental run, the context for the first 2 min is DayTime, changing to NightTime in the next 3 min and back to DayTime until completion. During NightTime, mote 1 transmits data with the exception of light readings. For a total of 207 recorded readings, only 5675 bytes ($41\times21 + 166\times29$) have been transmitted by the motes utilizing the framework, whereas a control experiment for 7 min would require a transmission of 6003 bytes (207×29) under the same conditions. This has shown that effective change of context states will reduce the load of messages that need to be sent, where redundant packets could be omitted, and will do so more significantly when the sensor network has to operate for a longer period of time.

Experiment 4 — sleep mote experiment: Another aspect to examine when using a context-aware framework is the possibility of putting motes to sleep over long periods of time for optimum energy savings when no activity is expected. In this experiment, the context-triggered action extends to both the monitoring mote and the controlled mote. The context action is triggered when the monitoring mote (1) detects a context change from DayTime to NightTime and vice versa. During NightTime, the motes will be put to sleep for an extended period of time until DayTime is expected.

The readings obtained from the experiments have in effect demonstrated that the architecture is feasible from the implementation perspective involving a real sensor toolkit. However, we also note that an implementation of the framework needs to be sufficiently efficient in detecting context changes and translating such changes into commands for controlling sensors, and the context changes should not occur too frequently — the scenario is such that a context state remains unchanged long enough to make it worthwhile changing sensor power states. This implies that if context changes occur too frequently, then the framework might not be able to respond to such changes in time.

7.4 SUMMARY

We have discussed the idea of sensors being aware of their own context or circumstances or the context of other sensors and exploiting that awareness for energy-saving behavior. We have illustrated these ideas by presenting a framework and a prototype based on a scenario. Although many existing approaches have looked at improvements in sensor hardware with restricted applications, the approach here emphasizes a software model for energy benefits in sensor networks.

There are many ways to exploit the concept of the context-aware sensor. For instance, sensors (with the right capability) can move toward sensed sources or toward energy sources (e.g., move from shaded areas to areas under strong sunlight, or move toward sources of heat or toward vibrations[2]).

ACKNOWLEDGMENT

This chapter contains portions reprinted, with permission, from Chong, S.K., Krishnaswamy, S., and Loke, S.W., A context-aware approach to conserving energy in wireless sensor networks, *Proceedings of the First International Workshop on Sensor Networks and Systems for Pervasive Computing (PerSeNS 2005) held in conjunction with the Third International IEEE Conference on Pervasive Computing (PerCom 2005)*, Kauai, HI, March 8–12, IEEE Press, pp. 401–405. *This chapter also contains portions from an unpublished draft paper authored with Suan Khai Chong, used with permission.*

---

[2] Kinetic energy can be harvested for sensors — http://www.perpetuum.co.uk/.

REFERENCES

Akyldiz, I.F., Su, W., Sankarasubramaniam, Y., and Cayirici, E., A survey on sensor networks, *IEEE Communications Magazine* 40, 2002, pp. 102–114.

Cardell-Oliver, R., Smettem, K., Kranz, M., and Mayer, K., A reactive soil moisture sensor network: design and field evaluation, *International Journal of Distributed Sensor Networks* 1(2), 149–162, April–June 2005.

Chong, S.K., Krishnaswamy, S., and Loke, S.W., A context-aware approach to conserving energy in wireless sensor networks, *Proceedings of the 1st International Workshop on Sensor Networks and Systems for Pervasive Computing (PerSeNS 2005) held in conjunction with the Third International IEEE Conference on Pervasive Computing (PerCom 2005)*, Kauai, HI, March 8–12, IEEE Press, pp. 401–405.

Clarkson, B.P., Life Patterns: Structure from Wearable Sensors, Ph.D. thesis, School of Architecture and Planning, Massachusetts Institute of Technology, Cambridge, MA, 2002.

Elnahrawy, E. and Nath, B., Context-aware sensors, *Proceedings of the 1st European Workshop on Wireless Sensor Networks*, 2004, pp. 77–93.

Huaifeng, Q. and Xingshe, Z., Context-aware sensornet, *Proceedings of the 3rd International Workshop on Middleware for Pervasive and Ad-Hoc Computing*, Grenoble, France, 2005, ACM Press, pp. 1–7.

Mainwaring, A., Polastre, J., Szewczyk, R., Culler, D., and Anderson, J., Wireless sensor networks for habitat monitoring, *Proceedings of the ACM International Workshop on Wireless Sensor Networks and Applications*, Atlanta, GA, 2002, pp. 88–97.

8

CONTEXT-AWARE SECURITY

Traditional security models help to ensure integrity, nonrepudiation, and confidentiality of information. Security seems inherently contextual and relative to the environment and circumstances; what is secure in one situation is considered insecure in another; information is insecure if it is viewed not by the right people, not at the right time, and not in the right circumstances. With mobile and pervasive computing developments, information can be accessed in more ways and in more places than ever before and in forms not previously possible. Moreover, the context in which such information is viewed or used can change given the agility and mobility made possible by new devices and wireless networking technologies. Also, as noted in Chapter 2, advancement in sensing technologies have narrowed the gap between the physical and virtual world, so that the computer can have a better picture of the physical world via advanced, and increasingly widespread, sensor technology and sensor information processing techniques. With computers having the ability to acquire context information of people and things and recognize situations, the opportunity becomes available for computers to help manage secure information with its own better knowledge of the physical world.

Context-aware computing can influence security models in at least three ways:

■ Finer-grained security: Fine-grained security models have been developed based on the notion of context awareness. For example, an appropriate security measure weaker than traditional security models yet strong enough for the situation at hand can be applied. Context might also be a means in which individuals are grouped for the purposes of security decisions; only those people who have passed through this room can view this information. Alternatively, context can be used to enhance existing security measures. A scene sometimes seen in movies is how retina scanning or handprint scanning might be used as biometrics security for entry through a door and how (quite gruesome it might be) an eyeball or hand might be used from a dead person to unlock a door. Combined with context sensitivity, such scenarios might be made impossible; in addition to the biometrics, a camera might be used to partially recognize the person in front of the door, or weight sensors can be employed on the doormat in front of the door, as additional context for access control. Alternatively, vital signs from the body of a person might be needed in addition to the biometrics data.

■ Adaptable security levels: With awareness of the context of entities, flexible security models can be achieved; enabling security levels can be increased or decreased not just based on the identities of people but the situation they are currently in. For example, to access the same information, in some situations, a driver's license might be adequate, and, in others, the combination of the right location, time, people nearby, and device identifiers might instead be adequate. More generally, one set of contexts and credentials might be thought to be just as adequate as another set of contexts and credentials. Alternatively, to access the same information, the security level can change depending on the circumstances. For example, given some information to be protected, if this information is accessed at a certain time and place or, more broadly, in a certain situation, then the security level can be lower than if the information was accessed at a different time and place or a different situation. More generally, the security level might be adjusted based on context. Another method to use context is to present partial information; full disclosure might only be available in certain situations, and, in other situations, only partial disclosure is permitted.

■ Increased traceability: It may be possible to allow access to secure information even when a person does not have the full credentials required at the time, provided the person is willing for the transaction to be comprehensively recorded. For instance, a person is allowed to view certain information without a required password, provided the person is willing to have his or her picture taken and leave behind thumb prints, with the context of location and time and witnesses of the event recorded. In general, context can help to improve traceability and enrich auditing, and perhaps it can be traded for changes in required credentials or security levels.

Mostéfaoui (2004) defines the term *context-based security* as follows:

Context-based security supports the reconfiguration of the security infrastructure according to the situation of use. This reconfiguration is governed by the current context, formally called a security context.

Security context refers to the information collected about the user's environment that is applicable to the security system at hand. Dulay (2004) presents a comprehensive array of the types of context information that would be useful in security applications. They include the following:

- Current state: the user's current location, time, activity, people nearby, physiological state, available services, network connectivity, etc. We have already discussed the use of such context information, and they are indeed applicable to enhance security.
- User preferences and relationships: includes recommendations from people. This type of context information is interesting as it involves invoking personal and social information in making security decisions. For example, in emergency situations, an authenticated family member might be able to access the information regarding an injured person or be permitted to access vital information.
- History: accumulated wisdom. Historical information might be of use in relation to trust based on previous outcomes; for example, in decisions about permitting access, although this is sometimes difficult to capture.

The rest of this chapter explores state-of-the-art security models that utilize context, showing the potential of the technology.

8.1 TRADITIONAL SECURITY ISSUES AND MODELS

We first briefly review background concepts in security. Further details can be obtained from other sources such as Anderson (2001). Traditional security models aim to provide confidentiality, integrity, and availability of information. Confidentiality involves ensuring that information is only shared or accessed by authorized parties under specific conditions; i.e., it relates to the privacy concerns of an asset. Authentication, which is the process of ensuring that the identity declared is indeed the true identity, is important in enabling access to the right parties. Integrity involves ensuring that the information can be trusted and has not been tampered with (e.g., during transmission, etc.) and also that, if the information has indeed been modified, what to do under such circumstances (e.g., trace the source of the modifications and take required action). Nonrepudiation involves ensuring that senders of information, cannot deny having sent the information, and receivers cannot deny having received the information. Availability relates to the systems handling the information being accessible by authorized parties in the required circumstances. Denial-of-service, for example, is one form of attack on a system, which aims to reduce, or cause the complete loss of, access to resources or services, or render inaccessible the resource or service to legitimate parties by overloading the system providing the resources or services.

An important concept in security systems is the security policy, which describes valuable (typically information-based) assets to be protected and specifies security responsibilities. Such policies might be described informally or in a formal mathematical language.

8.2 CONTEXT-AWARE SECURITY SYSTEMS

We review several examples of systems and different ways to represent security policies that take context into account.

8.2.1 Examples

We will start with an example of using context to enhance a traditional security application. Logging into a computer system with username and password is perhaps one of the most common authentication mechanisms for using the computer. Bardram et al. (2003) present a more comprehensive mechanism to enable what is called proximity login, i.e., login by approaching the computer physically (with an authentication token), which works as follows:

1. Uses a physical token (e.g., smart card) for gesturing and as the cryptographic basis for authentication.
2. Uses a context-aware system to verify the location of the person and logs the person out when he or she leaves (i.e., is found as being not within the same location as) the computer.
3. Incorporates a fallback mechanism; if the positioning infrastructure fails to determine the user's location, then the user is requested to enter his or her password to log in.

Location information is acquired via long-range RFID monitors reading passive RFID tags, and by WLAN monitors that can tell the cell-based location of networked devices. Given errors with the location technology, a probability is assigned to the estimated location of the person, and only if the probability is below a specified threshold is the person required to provide a password. A user can log into a computer via the smart card provided the system perceives with high probability that the user is indeed close enough to the device (e.g., in the same room); i.e., it is not enough to steal the smart card to log into the computer without the user being adequately close to the computer.

The aforementioned system demonstrates the use of context to provide (1) additional security (in addition to the smart card), and (2) alternative security (apart from using a password). Generally, in security frameworks that support a larger variety of scenarios, a security policy coupled with a context-awareness infrastructure would be useful, to specify security behaviors in different settings.

An example of using context to provide flexible security for wireless network environments is described by Hager (2004). Context is used to choose the right security protocol or algorithms. Based on such context, a decision engine computes the appropriate protocol or algorithm. Six types of context are used:

- Security level: The security level is set by the user. A high security level would cause a stronger security algorithm to be used for encrypting data.

- Energy: The energy level of a device on which security algorithms are to be employed is measured. First, a check is made to see if the device is connected to the power supply; if not, the remaining battery life of the device is measured. Low battery life leads to the selection of a more energy-efficient security algorithm.

- Location: Location can determine the level of required security. Possible locations used in the system are Home, Office, Lab, and Unknown. Transmission between devices both at home might not require such a high level of security. Wireless data transmission from a device at home might not require such a high security level as from a bank. Unknown locations might be deemed insecure.

- Communications: Parameters concerning communication include throughput, link capacity, quality of service, and signal strength. Because data rates will also drop if the signal strength is weak, then, with high bandwidth communications with strong signal strength, stronger security algorithms that are less efficient in terms of data exchanges might be employed.

- Object size: Larger files require more time to encrypt and decrypt, and so more efficient encryption algorithms might be employed for larger objects.

- User interactions: The more times two users have interacted with each other and in more sessions, the lower is the security level deemed required, so that algorithms involving rapid communication with reduced energy consumption is favored.

A value for each of the aforementioned parameters is normalized by mapping to a context metric value. The combined set of context metric values is then aggregated and weighted according to which context is more important, and a security algorithm is chosen.

As a third example, from Shankar and Balfanz (2002), we consider how context can be used to group entities to form secure associations among the entities. Such a secure association then provides a basis for secure ad hoc communication among the entities. A key concept is the *context view*. A context view is part of an entity's context that an application is interested in. The idea is that software components called context view providers observe devices and their contexts and generate context views. A security software component processes a collection of context views to find associations among these views. For example, it could find that several devices are located in the same room, and, by virtue of similar context, such devices are grouped together. Examples of applications given include the following:

1. Instant secure e-mail: An e-mail sent to room123@mycorp.com will reach all members currently located in conference room 123.
2. Secure file-sharing applications: Members in context-view associations can securely share files or documents.
3. Instant secure groupware: Groupware, including Web conferencing tools, can be enabled within members of a secure group.

This approach provides associations or grouping by context, which might change, and so, such ad hoc grouping provides secure context for short-term or transient interactions.

Each of the aforementioned systems enhances and enables a specific security application. More general frameworks have been proposed in which users can define context-aware policies tailored to different scenarios and applications.

8.2.2 Context-Aware Policies

This subsection provides illustrations of context-aware policies and architectures of context-aware security systems.

A context-aware security policy incorporates contextual information of entities or situations within the rules of the policy. Context can be incorporated into policies in a number of ways, from augmenting conditions for access to particular information to triggering change in the set of applicable security policies.

There are a number of different formalisms for expressing context-aware security policies. We review several examples in the following text, including contextual graphs, logic, and role-based XML languages.

8.2.2.1 Contextual Graphs

The contextual graph is similar to decision trees and provides a graphical formalism for expressing security policies incorporating context information (Brezillon and Mostefaoui, 2004). A contextual graph is a directed acyclic graph with a unique input, a unique output, and an organization of nodes connected by arcs, where each node can be an action, an indication of the contextual information, or a recombinant node. A path through the graph accumulates contextual information leading to a decision on security actions. Figure 8.1 gives an example of a contextual graph. Such a graph shows the security policy for a healthcare records distributed system.

The security action is indicated by the square boxes. The action Authentication Method P is done for surgeons, followed by other security actions (involving cryptographic protocols), depending on further context (i.e., the location of the surgeon). For example, if the role is nurse and the current location of the nurse is loc1, then Authentication Method Q is used, followed by Cryptographic Protocol Y, as dictated by the particular path through the graph. Such contextual graphs provide a visual representation of rules and allows decisions and their security consequences to be traced out easily. R1, R2, and R3 are recombinant nodes corresponding to decision points (involving context) C1, C2, and C3.

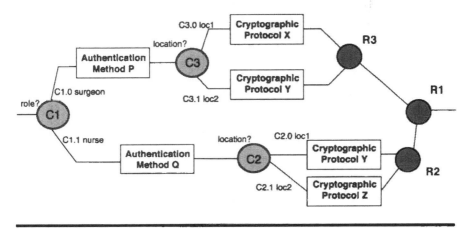

Figure 8.1 Example of contextual graph.

8.2.2.2 Logic

Logic can be used to represent security policies in a way that is human-readable yet precise enough to be machine-processable, allowing reasoning to be carried out. For example, the Gaia environment has a service called Cerberus, which integrates context awareness and reasoning with security policies (Al-Muhtadi et al., 2003). Context information is represented using predicates such as the following:

```
Location(Bob, entering, room 2401)
Temperature(room 3231, "=", 98F)
Time(New York, "<", 12:00 01/01/01)
```

Boolean operators can be used to form more complex expressions such as Location(Tom, entering, room 3000) ∧ SocialActivity(room 3000, meeting), which refers to the context that Tom is entering room 3000 and that there is a meeting going on in that room. In the style of first-order logic, one could also use quantifiers over locations, people, and other variables in predicates.

Different strengths of authentication are associated with different confidence levels. Modeling of the different strengths of authentication is useful given the flexibilities afforded by the use of context in security policies. It is also useful in catering to different security technologies that can be used. Facts such as the following represent confidence levels associated with different technologies:

```
ConfidenceLevel(smart_badge, 30%)
ConfidenceLevel(smart_card, 80%)
```

Then, according to whether a principal P has been authenticated according to technology, facts such as the following are added to the system:

```
Authenticated(P, password)
Authenticated(P, fingerprint)
Authenticated(P, retinascan)
Authenticated(P, mobilephoneSIM)
```

Confidence values can then be computed using a rule such as the following:

```
ConfidenceValue(P,V):-
    ∃device X(Authenticated(P,X)^ ConfidenceLevel(X,V))
```

The operator ":-" means "if." Access control decisions are then represented using a rule such as the following:

```
CanAccess(P, dataProjector):-
    ∃_number V(ConfidenceValue(P,V)^ V>60%)
```

This rule states that the data projector (in a room) can only be accessed if the confidence value of P's authentication is above 60%. Note that one could extend the rule to accommodate context information easily using a rule such as this (which also requires that P be located in the seminar room):

```
CanAccess(P, dataProjector):-
    ∃_number V(ConfidenceValue(P,V)^ V>60%)^
    Location(P, seminarRoom).
```

The inference engine can be used to answer queries such as whether a principal named John can access the data projector:

```
?- CanAccess(john, dataProjector)
```

8.2.2.3 Roles

Role-based access control policies have been extended with context information. For example, a graphical editor based on the Generalized Policy Definition Language (GPDL) is presented in Covington et al. (2001) for representing policies in XML format, which uses the role as the key concept. Each policy contains a subject role (the subject on which the policy is to apply), an object role (the object to which the policy applies), an action (on the object), an environment role (the context of the action), and the permissions (whether the action is to be permitted or not). For example, the following policy specifies that a child is denied all actions on dangerous appliances during working hours:

```
<GRBAC_TABLES>
    <POLICY>
        <SROLE> Child </SROLE>
        <OROLE> Dangerous Appliance </OROLE>
        <ACTION> ALL </ACTION>
        <EROLE> Working Hours </EROLE>
        <PERMS> Deny </PERMS>
    </POLICY>
</GRBAC_TABLES>
```

The system that processes such policies determines what entities occupy which roles in deciding access control actions. The environment role represents the context of the action (the aforementioned example considers only time but can be more elaborate). Context information is obtained in the system via a context management system built using the Context Toolkit.[1]

Also using the notion of roles are the context-aware policies in Tripathi et al. (2004), given likewise in XML format. We provide the following example as given in the paper but in a simpler format than XML:

```
ActivityTemplate RoomController(ObjectType Room room) {
    Role LightManager {
        Reaction SwitchOnLight {
            Precondition #(room.presentUsers()) > 0
            Action room.light.switchOn() }
        Reaction SwitchOffLight {
            Precondition #(room.presentUsers()) = 0
            Action room.light.switchOff() }
        Reaction Dim {
            Precondition #(room.presentUsers()) > 0
                & room.projector.on()
        Action room.light.setLevel(LOW) }
}}
```

This policy specifies actions on the room lights according to the number of people present in the room — turn the lights off if no one is present and turn the lights on if at least one person is present. There is also a rule specifying that the lights be dimmed if the projector is on and there are people in the room. Other policies more closely related to security can be defined, such as a policy that specifies that a room door be locked automatically every day whenever no one is in the room and at specified times. The LightManager's role is to perform the action according to conditions stated.

Recent work by Gomez et al. (2005) extends the XACML (eXtensible access control markup language) for security polices in the context of a flexible patient record access application. XACML supports the role-based access control model. The extended XACML language allows the user to specify context attributes that involve not only user location in terms of

---

[1] More information on the Context Toolkit is available at http://www.cs.cmu.edu/~anind/context.html.

latitude, longitude, distance (from a given point), and what an object is close to (i.e., a proximity condition), but also health-related contexts such as the health status of the patient (based on data about heart rate, blood pressure, and temperature) and whether there is an emergency. The system uses GPS receivers for location and sensors to acquire patient data. Hence, during emergency situations in which the physician is close to the patient and the patient's health is in danger, access to the patient's otherwise confidential medical records is permitted. A demonstrator for an E-health scenario was developed.

The architectures of systems that process the aforementioned policies have the common feature of a component to interpret and enact the policies, and a second component to execute security actions, in consultation with a third component, which acquires the context information of entities. Indeed, these components can be decoupled in the sense of the three subsystem architecture described in Chapter 2. This means that different infrastructures can be "plugged in" to provide context awareness for security purposes. For example, the Context Toolkit was used by Covington et al. (2001) as noted earlier, and the Trust Context Spaces system of Robinson and Beigl (2003) uses the Smart-Its sensors system to provide context information for generating cryptographic keys.

There is a range of policy languages as already described, for explicitly representing security measures that incorporate context.

8.3 FROM CONTEXT-AWARE SECURITY TO CONTEXT-AWARE SAFETY

Security and safety consider different issues but have adequate similarity so that the aforementioned work can be applied to safety applications as well. According to the *The American Heritage® Dictionary of the English Language* (fourth edition), the word *security* means "freedom from risk or danger; safety," and the word *safety* means "the condition of being safe; freedom from danger, risk, or injury," and also "a device designed to prevent accidents, as a lock on a firearm preventing accidental firing." Hence, these two words are very closely related.

We can consider a smart-home scenario from Shetty and Loke (2005) in which appliances in the kitchen can only be activated in the presence of an adult. A young child's position is sensed, and, if not collocated with an adult in the kitchen, the child will not be able to operate the stove. Again, if the temperature at the stove is getting higher or the smoke alarm detects some amount of smoke, heat from the stove can be automatically turned off. Hence, instead of mapping context to security levels or security actions, it is possible to map context to safety levels and safety-related actions. Other work such as that of Salim et al. (2005) attempts to sense situations on the road and warn drivers of impending dangers. For example, a driver can be warned that a high-speed car is approaching the same junction and shows no sign of stopping. Also, drivers can be alerted, or action taken, when the system detects via in-car sensors that they are drowsy, drunk, or tired. When context-aware security is applied to a home, then the line between safety and security blurs. The security system of the house can be adapted according to context. For example, security access to the doors and windows and other security mechanisms may be enhanced when the residents are not in the house or when the residents are asleep.

8.4 SUMMARY

We have reviewed new advantages that employing context awareness brings to security, including more flexible and fine-grained security models.

There are other related issues that we have not discussed in depth in this chapter. One is the use of context not so much for securing information as for securing context information, and related privacy issues. One would not like one's whereabouts to be continually tracked or publicly known even if that provides new flexibility in security or new mobile services. There have been recent works on protecting people's location information (Hengartner and Steenkiste, 2003) and on mechanisms to allow users to have more control over how their context information will be used (such as the work by Hong in his Ph.D. thesis as mentioned in Chapter 2).

Another issue relates to the use of context in security infrastructures. The context information itself has to be trustworthy and secure — false context can compromise security. Protecting the sensing and context gathering subsystems will be important so that false information is not acquired.

Security in pervasive computing environments is an emerging area of interest, which raises issues not typical in traditional distributed computing, such as vulnerabilities with wireless networking, introduction of new unknown devices to an environment, denial-of-service for openly available devices (e.g., sleep deprivation of limited battery power devices), and integrity violations of critical wireless sensor data through intercepts (Stajano, 2002; Nixon et al., 2004).

REFERENCES

Al-Muhtadi, J., Ranganathan, A., Campbell, R., and Mickunas, M.D., Cerberus: a context-aware security scheme for smart spaces, *Proceedings of the 1st IEEE Annual Conference on Pervasive Computing and Communications (PerCom 2003)*, Fort Worth, TX, 2003, pp. 489–496, available at http://www.cyberdudez.com/cerberus.pdf.

Andersson, R., *Security Engineering: A Guide to Building Dependable Distributed Systems*, John Wiley & Sons, New York, 2001.

Bardram, J.E., Kjaer, R.E., and Pedersen, M.O., Context-aware user authentication — supporting proximity-based login in pervasive computing, in Dey, A., McCarthy, J., and Schmidt, A., Eds., *Proceedings of Ubicomp 2003: Ubiquitous Computing*, Vol. 2864 of Lecture Notes in Computer Science, Seattle, WA, October 2003, Springer-Verlag pp. 107–123, available at http://www.daimi.au.dk/~bardram/docs/bardram.ubicomp2003.pdf.

Brezillon, P. and Mostefaoui, G.K., Context-based security policies: a new modeling approach, *Proceedings of the 2nd IEEE Annual Conference on Pervasive Computing and Communications Workshops (PERCOMW'04)*, 2004, IEEE Computer Society Press.

Covington, M.J., Long, W., Srinivasan, S., Dey, A.K., Ahamad, M., and Abowd, G.D., Securing context-aware applications using environment roles, *Proceedings of SACMAT'01*, 2001, ACM Press.

Dulay, N., Adaptive Context Aware Security, presentation at UK-UbiNET, May 2004, available at http://www-dse.doc.ic.ac.uk/Projects/UbiNet/ws2004/Slides/Dulay.pdf.

Gomez, L., Moraru, L., Simplot-Ryl, D., and Wrona, K., Using sensor and location information for context-aware access control, *Proceedings of EUROCON 2005*, Belgrade, November 2005, available at http://www.lifl.fr/RD2P/uploads/Papers/gomez-eurocon-05.pdf.

Hager, C.T., Context Aware and Adaptive Security for Wireless Networks, Ph.D. thesis, Faculty of the Virginia Polytechnic Institute and State University, 2004.

Hengartner, U. and Steenkiste, P., Protecting people location information, *Proceedings of the 1st International Conference on Security in Pervasive Computing*, March 2003, pp. 25–38.

Mostéfaoui, G.K., Towards a Conceptual and Software Framework for Integrating Context-Based Security in Pervasive Environments, Ph.D. thesis, Universite Pierre et Marie Curie, Paris VI, 2004.

Nixon, P., Wagealla, W., English, C., and Terzis, S., Security, Privacy and Trust Issues in Smart Environments, Computer and Information Sciences, Smartlab Technical Report (Smartlab-2004-01), 2004, available at http://smartlab.cis.strath.ac.uk/Publications/techreports/SPTPaperFinal.pdf.

Robinson, P. and Beigl, M., Trust context spaces: an infrastructure for pervasive security in context-aware environments, *Proceedings of the 1st International Conference on Security in Pervasive Computing*, Germany, 2003, Springer-Verlag, Lecture Notes in Computer Science 2802, 2004.

Salim, F.D., Krishnaswamy, S., Loke, S.W., and Rakotonirainy, A., Context-aware ubiquitous data mining based agent model for intersection safety, *Proceedings of the Embedded and Ubiquitous Computing Workshops*, 2005, pp. 61–70.

Shankar, N. and Balfanz, D., Enabling secure ad-hoc communication using context-aware security services, *Proceedings of the Workshop on Security in Ubiquitous Computing, at the Conference on Ubiquitous Computing*, 2002.

Shetty, P. and Loke, S.W., Context-based security (and safety) meta-policies for pervasive computing environments: the case of smart homes, *Proceedings of the CONTEXT-05 Workshop on Safety and Context*, de Lavalette, B.C. and Tijus, C., Eds., Paris, 2005.

Stajano, F., *Security for Ubiquitous Computing*, John Wiley & Sons, U.S.A., 2002.

Tripathi, A., Ahmed, T., Kulkarni, D., Kumar, R., and Kashiramka, K., Context-based secure resource access in pervasive computing environments, *Proceedings of the International Workshop on Pervasive Computing and Communications Security (IEEE PerSec'04)*, 2004, IEEE Computer Society Press.

9

CONTEXT AWARENESS AND MIRROR-WORLD MODELS

This chapter discusses an idea for implementing awareness. The idea is that a virtual model of the real world is constructed and some amount of synchronization between the virtual model and the real world is managed via sensors and context awareness. The form of the virtual model, the details captured, how up-to-date the model is, and the degree of correspondence to the real world will depend on the specific application. We consider several such models in the following text.

We first consider the idea of mirror worlds introduced by Gelernter (1993), and then discuss a project called Nexus, which aims to construct and employ virtual models that reflect aspects of the real world augmented with additional virtual objects. We then note the parallel between these mirror virtual models and 3-D virtual worlds used for multi-user dimension (MUD[1]) games. We also discuss an example of how ideas developed for virtual environments can be employed in the physical world once we observe that the virtual environment can mirror the physical world or, perhaps, an abstraction of it. We then review the notion of smart spaces or smart environments and briefly discuss ontologies for building mirror worlds.

9.1 GELERNTER'S MIRROR WORLDS

A mirror world is effectively a virtual model or counterpart of some part of the physical world, such as a city mirror world or a hospital mirror world, which contains detailed descriptions of a city (e.g., maps, etc.) or

---

[1] http://en.wikipedia.org/wiki/MUD.

a hospital, respectively. One can browse deeply within mirror worlds to different levels of detail; such mirror worlds aim to reflect the physical reality in real-time. To quote from Gelernter (1993):

> ... you flip channels until you find the Mirror World of your choice, and then you see a picture. Capturing the structure and present status of an entire company, university, hospital, city, or whatever in a single (obviously elliptical, high level) sketch is a hard but solvable research problem. The picture changes subtly as you watch, mirroring changes in the world outside.

One can also interact with software agents in mirror worlds or with other visitors (perhaps real people in the physical world), or insert new agents; according to Gelernter (1993):

> You meet your software agents and other Mirror World visitors along the way ... you choose to ask questions or plant new agents ...

To accurately reflect the physical world, a mirror world is fed with information from data-gathering or monitoring equipment, perhaps some combination of sensors, and sensor data filtering and processing:

> A Mirror World is an ocean of information, fed by many data streams. Some streams represent hand-entry of data ... Others are fed by automatic data-gathering and monitoring equipment, like ... weather-monitoring equipment, or traffic-volume sensors installed in roadways.

Many such mirror worlds might be constructed by different people and integrated into a whole, and accessed by many people via different computers simultaneously.

It is interesting that mirror worlds were envisioned by Gelernter before the World Wide Web was invented or became as large as it is today. In some sense, the Web reflects, albeit weakly, the physical world — from personal home pages as counterparts of individuals to university Web pages corresponding to universities (and, similarly, faculty home pages corresponding to faculties and department home pages corresponding to departments), and from businesses, often with their own home pages, to places with theirs. However, keeping these Web pages up to date, often done manually, is tedious work. It is also interesting that mirror worlds were envisioned by Gelernter before sensor networks gained the popularity and enjoyed the developments seen today. The idea of feeding data streams into a computational structure so that the data coming in is reflected in changes to some high-level model is used to help interpret the data and make it more comprehensible, in the same way that a context-aware system would aggregate or process sensory data into high-level context meaningful to an application.

Holzhauer (2005) proposes an integration of modeling and simulation with wireless sensor networks similar to the idea of mirror worlds though not browsable or visual. It helps to reduce data transmission, which costs energy, and therefore saves battery power on the sensors. The idea is that the simulation provides forecast of data values, and only deviations from the forecasted values are transmitted; where there is no transmission, it is assumed that the forecasted value is correct. This can reduce the frequency of transmitting sensor data. The model in this case is a mathematical model for predicting sensor data values.

9.2 NEXUS

What can be viewed as a serious attempt at constructing mirror worlds is the Nexus[2] project. The project clearly distinguishes three layers of abstraction: the physical world, the augmented-world model, and information spaces. The augmented world refers to a model that mirrors aspects of the physical world and yet contains virtual objects that "augment" the physical world. The Nexus framework provides the Augmented World Modeling Language (AWML) and the Augmented World Query Language (AWQL) (Nicklas and Mitschang, 2004; Lehmann et al., 2004), which can be used to describe objects and relations between objects in an augmented-world model, including objects that correspond to real-world objects and have position and extent, virtual informational objects such as objects situated on Web servers, and objects containing navigational information. Real sensors are modeled as special spatial objects. Virtual sensors that combine several real sensors and provide synthesized or processed information can also be defined. Associations can be specified between augmented-world objects and real-world objects (e.g., associating Web pages with particular real-world objects such as exhibits of a museum). Relations supported in the Nexus model include "part of," "sticks on," "belongs to," and "held by." Applications using Nexus include a Museum Guide, Navigation Tool, and Virtual Scavenger Hunt.

Based on such augmented-world models, event-based applications can be built, where events can be expressed in a high-level model at an abstraction level that corresponds to real-world everyday language descriptions (e.g., the user entering a particular area, passing particular objects,

---

[2] http://www.nexus.uni-stuttgart.de/en/overview/vision/index.html.

or near another person) (Bauer and Rothermel, 2004). In the case of uncertainties inherent in sensors, threshold probabilities can be employed; when the probability of the event is above a certain value, the event is deemed to really occur.

One cannot help but note the overlap with the smart virtual counterparts project mentioned in Chapter 4, in which real-world objects can be associated with its virtual counterparts via RFID tagging. However, augmented-world models allow more comprehensive spatial modeling. The notion of augmented reality,[3] which superimposes information directly over the physical environment viewed via a headset and a wearable computer, is also interesting and bears some resemblance to physical worlds augmented with virtual objects, of which an augmented-world model provides an explicit representation. In augmented-reality work, one can look at a real-world object via special head-mounted glasses and have information about the object displayed on the glasses superimposed over the real-world object. Such superimposition is similar to how virtual information objects (e.g., virtual stick E-notes) might be superimposed on maps of real-world places.

9.3 VIRTUAL WORLDS, VIRTUAL ENVIRONMENTS

Virtual worlds have been employed in many computer games, social interaction, and entertainment applications. MUDs comprise 3-D worlds in which persons or their representatives called *avatars* might dwell and meet other avatars.[4] A MOO is a textual view of such virtual worlds but with text that paints a picture of people, things, and places, such as the following:

> You are standing on the sandy banks of a small river. A small wooden bridge spans the river, leading to the jungle on the other side. You see a shiny spot in the midst of the green right in front of you ....

Such virtual worlds typically have no correspondence to real-world places but are constructed with spatial relationships and objects so as to provide adequate engaging realism.

One can note the similarities between such virtual worlds and mirror worlds, except that mirror worlds do attempt to correspond to some real-world place, even if augmenting it with virtual objects, whereas virtual worlds for games and entertainment applications generally do not. This

---

[3] http://www.augmented-reality.org/.

[4] An atlas of virtual worlds can be found at http://www.cybergeography.org/atlas/muds_vw.html.

similarity can be exploited to create sophisticated virtual worlds that mirror physical worlds in the spirit of the Nexus' augmented-world models. For example, real world movements of people through physical spaces can be mirrored in the virtual world by the movement of their corresponding avatars in virtual spaces — imagine a virtual world mirroring a stadium where a soccer match is going on. Events that happen in the physical world are mirrored via changes in the virtual world. Beyond merely reflecting the real world, the converse might be possible in some cases. Actions in the virtual world (e.g., on objects in the virtual world) might cause actions in the physical world (e.g., invoking commands on a television representative in a virtual world causes the actual television in the physical world to be turned on), or moving a robot in the virtual world causes corresponding movements of a physical robot in the real world. Issuing commands or queries to objects in the virtual world results in commands or queries being actually issued to objects in the physical world. In fact, this relates to the virtual reality concept, whereby a user has the experience of being embedded in an environment that is simulated in a computer. A variety of sensory inputs increase the sense of realism (e.g., via haptic feedback) and allow interaction with objects in the virtual environment. With mirror worlds, interacting with the virtual world is interacting with the physical world being mirrored.

Indeed, virtual-world counterparts of physical worlds allow augmentations to the physical world in new and interesting ways. Inherent in Nexus' AWML is the notion of augmenting physical worlds with informational and computational objects. In the following text, we consider another example of enriching the physical world via computational notions first introduced in virtual environments.

9.3.1 Aura, Focus, and Nimbus: Virtual Objects and Real-World Objects

We consider an example of a model for mutual awareness of devices or physical artifacts inspired by the spatial model of mutual awareness used in MASSIVE (Greenhalgh and Benford, 1995) for virtual environments, in which entities in a virtual environment are made aware (or unaware, as is the case) of one another via a model involving concepts such as *nimbus*, *focus*, and *aura* surrounding entities. Benford et al. (1993) stated that communication between objects within the virtual environments is a necessity, and a way to achieve this is by allowing each virtual object to have its own aura. As stated by Ferscha et al. (2004), aura is a subtle sensory stimulus of "attraction" that transmits "signals of attraction" governed by the "laws of attraction." Benford et al. (1994) defined aura as a "subspace which effectively bounds the presence of an object within a

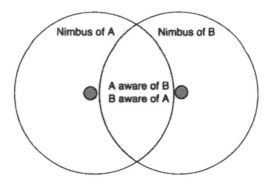

Figure 9.1 Nimbuses of two devices. (From Benford, S. et al., Managing mutual awareness in collaborative virtual environments, *Proceedings of VRST '94*, Singapore, 1994, ACM Press, pp. 223–236.)

given medium which acts as an enabler of potential interaction." It means that each virtual object has a territory of virtual space that surrounds the object. Information exchange or establishment of connection between the two objects occurs when these territories cross over and thus interaction between these objects is made possible within the virtual space. Aura is just an elementary mechanism or stimulus for interactive behaviors for these objects. The size or shape of these auras varies independently as each object has its own criteria to meet. The mechanism for interaction of objects is made possible with auras, the objects themselves now being responsible for controlling these interactions. This process of controlling the interaction is achievable by having degrees in the level of awareness between these objects. It is with this measurement of focus and nimbus that an object's interaction is redirected toward another object and one can adjust the extent to which an object is aware of another (Benford et al., 1994; Rodden, 1996). The definition taken from Benford et al. (1994) states that:

- The more an object is within your *focus*, the more aware you are of it.
- The more an object is within your *nimbus*, the more aware it is of you.

In short, Rodden (1996) stated that focus is a function that allows an object to be mapped into a position within a space, whereas nimbus is a function that maps objects into the shared space and returns the position. By knowing the degree of focus and nimbus between objects, the level of awareness among them can be determined.

We can bring such concepts over to devices and everyday artifacts so that we can speak of the nimbus, focus, and aura of physical artifacts. We can conceive of physical artifacts having an aura surrounding them, each having its own nimbus and focus. A general model of mutual awareness can be adapted for different context environments, providing individual devices some control over their awareness toward other devices and the device's information (while preserving privacy). We can implement the concepts from the spatial model of awareness in a number of ways. For example, the aura of each device is assumed to be the area within the range of communication. Thus, given the focus and nimbus of each device, the level of awareness can be used to control and behave differently under different programmable conditions. For example, consider a device with Bluetooth networking capability. Given the Bluetooth range of 10 m, one can roughly define the aura (or the focus and nimbus) of a device as a sphere of 10 m radius. Alternatively, if another technology is used, say, RFID tags with several meters of range, one can implement the focus and nimbus of tagged objects in terms of these RFID parameters. Each device or artifact can adjust its own focus size (i.e., its level of awareness of others) and its own nimbus size (i.e., its level of concealment from others).

Given a collection of such devices, each with its own level of awareness and concealment, interesting behaviors can emerge, with certain devices able to interact with others provided the focus and nimbus permit it. A simple formalization of these notions can take the form of rules of awareness and concealment represented using formulas, depicting situations in which devices are aware of each other.

Let a be the level of awareness. Let c be the level of concealment. The minimum value of a and c is 1. The maximum value of a and c must be equal (as prescribed by the application designer, n). We define the following:

- Definition A: For device x to be aware of device y, device x's a value must be higher then or equal to device y's c value, i.e., $a(x) \geq c(y)$.

  When this condition is true, device x is able to detect device y.
- Definition B: For device x not to be aware of device y, it is the inverse of the condition, where, $c(x) > a(y)$.

Based on these definitions, we can derive propositions that represent various mutual awareness situations. We give five examples, i.e., proposition 1 to proposition 5.

Proposition 1 (Mutual Awareness)

Let $x, y \in$ Devices; device x and device y are aware each other if and only if $a(x) \geq c(y)$ and $a(y) \geq c(x)$.

Proof: By definition A of awareness.

When more devices are added, then, all devices are aware of each other if and only if

$\forall\ d_1, d_2 \in D$, $a(d_1) \geq c(d_2)$ where D is the set of all devices.

Proposition 2 (Mutually Unaware)

Let x, y ∈ Devices; device x and device y are not aware of each other if and only if c(x) > a(y) and c(y) > a(x).

Proof: By definition B of awareness.

Proposition 3 (Asymmetric Awareness)

Let x, y ∈ Devices; device x is aware of device y, and device y is not aware of device x. Then, a(x) ≥ c(y) and c(x) > a(y)

Proof: From Definition A, device x is aware of device y if and only if $a(x) \geq c(y)$.

From Definition B, device y is not aware of device x if and only if $c(x) > a(y)$.

Proposition 4 (Condition 1 for Transitive Awareness)

Let x, y, z ∈ Devices; device x is aware of device y, and device y is aware of device z, then device x is aware of device z whenever c(y) ≥ c(z).

Proof: Suppose device x is aware of device y, and device y is aware of device z. Then, from Definition A, $a(x) \geq c(y)$ and $a(y) \geq c(z)$. But if $c(y) \geq c(z)$, then $a(x) \geq c(y) \geq c(z)$; i.e., $a(x) \geq c(z)$. Hence, device x is aware of device z.

When more devices are added, suppose x_1 is aware of x_2, x_2 is aware of x_3, x_3 is aware of x_4, x_4 is aware of x_5, ..., x_{n-1} is aware of x_n.

Then, x_1 is aware of x_n whenever $c(x_j) \geq c(x_n)$ for all $j \in \{2, ..., n - 1\}$.

Proposition 5 (Condition 2 for Transitive Awareness)

Let x, y, z ∈ Devices; device x is aware of device y, and device y is aware of device z, then device x is aware of device z whenever c(y) ≥ a(y).

Proof: Suppose device x is aware of device y, and device y is aware of device z. Then, from Definition A, $a(x) \geq c(y)$ and $a(y) \geq c(z)$. But if $c(y) \geq a(y)$, then $a(x) \geq c(y) \geq a(y) \geq c(z)$; i.e., $a(x) \geq c(z)$. Hence, device x is aware of device z.

When more devices are added, suppose x_1 is aware of x_2, x_2 is aware of x_3, x_3 is aware of x_4, x_4 is aware of x_5, ..., x_{n-1} is aware of x_n.

Then x_1 is aware of x_n whenever $c(x_j) \geq a(x_j)$ for all $j \in \{2, ..., n - 1\}$.

The implication of these statements is that as long as each device maintains $c(x_j) \geq c(x_n)$ or $c(x_j) \geq a(x_j)$, the awareness relationship will be transitive, regardless of the number of devices. Maintaining such a condition might be a "social rule" imposed on each device to have a transitively aware society of devices. Hence, the model allows different situations of mutual (non-)awareness to be represented. Adjustments to the level of mutual (non-)awareness can be done by adjusting the levels of awareness and concealment for each device or artifact (perhaps as dictated by the choice of the device or societal rules). So, a device can hide itself by setting its concealment level high or reduce the scope of what it can, or wants to, see or be aware of by decreasing its level of awareness.

Note that all the aforementioned computations about interartifact awareness happen within the virtual world (perhaps located in an "awareness" server that determines who or what should be aware of whom or what, based on the individual object's concealment and awareness levels). The results of such computations can be communicated to the objects so that behaviors can be physically manifested, thereby leading to interesting applications. For example, two teddy bears, if placed on the same sofa, might become mutually aware of each other and manifest this via fun exclamations and greetings, each teddy bear with embedded computers and the sofa acting as the awareness server, wirelessly networked to the computers in the teddy bears. Alternatively, the teddy bear may greet its owner in close proximity, i.e., on coming within its focus.

We have already seen in Chapter 4 that the idea of virtual counterparts of physical artifacts can be used to associate computational behaviors and informational objects with physical artifacts. The mirror world concept associates computational behaviors and informational objects with corresponding physical environments (parts of the physical world). The preceding model for mutual awareness shows how a simple model of focus and nimbus of objects within a computer (and computations governed by the rules already mentioned) can endow physical artifacts or devices with effective auras (and effective focus and nimbus), effective in the sense of such auras being useful for triggering interesting behaviors.

9.4 DIGITAL CITIES

The work on digital cities utilizes the metaphor of a city for a virtual world. A digital city is a 3-D reconstruction of a city, sometimes mirroring an actual city and sometimes an artificial city, similar to virtual worlds in games. We consider two examples of digital city projects.

Digital City Kyoto (Ishida, 2002) aims to mirror Kyoto, a city in Japan, itself. The digital mirror world of the city contains 2-D maps of the city as well as 3-D graphical models, constructed using 3DML[5]. Further, 2-D maps of the city contain hyperlinks to Web pages containing information

---

[5] http://www.flatland.com.

about various objects and buildings in the city. An interesting feature of Digital City Kyoto is that it aims to be live in the spirit of Gelernter's mirror worlds. Real-time sensors are deployed to gather information about features in Kyoto. For example, 300 sensors have been installed in Kyoto City to collect data about more than 600 city buses; each bus reports its location and routes every few minutes. Such real-time information can be useful to bus users and are accessed via mobile devices. Other real-time information noted are weather conditions and live video from the fire department. A virtual bus tour application through the digital city has also been built. Avatars can also populate the digital city. Digital City Kyoto was carried out within the scope of the Universal Design of Digital City project.[6]

A Canadian digital cities project[7] has plans to install environment sensors that read temperature, humidity, light levels, and proximity at short distances within the city of Montreal. Such sensor data is then relayed to Wi-Fi hot spots and then stored in databases. There are a large number of hot spots in Montreal[8] providing an effective networking backbone infrastructure to which such sensors can connect. With a digital model of Montreal, and knowing the position of the sensors, one can determine the context of the many streams of sensor data and interpret the data appropriately.

The aforementioned work has similarity with aspects of Gelernter's mirror world models. Such digital cities, however, are not easy to maintain (e.g., if new buildings arise or old buildings are pulled down, the 3-D models need to be updated). Also, there is scope for much greater depth to be added to such digital cities; one can imagine taking a virtual bus tour through the city and not able to enter the buildings displayed, but one could also have detailed models of the interiors of whole buildings, which users can then browse and explore. And, many more sensors can be added to the physical city to update the digital models at runtime. However, not all features of a digital model can be updated at runtime via sensory information. The problem of maintaining such complex models would then increase, unless it is a multiparty effort mirroring the way a large open distributed information system such as the Web grows. Creating large digital cities would require the effort and time of many and would be distributed over a large number of servers, analogous to the Web.

Such digital cities emphasize a visual 3-D model (akin to virtual worlds or virtual environments that also tend to emphasize the user experience

---

[6] http://www.digitalcity.jst.go.jp/home-e.html.
[7] http://www.digitalcitiesproject.net/.
[8] See http://www.ilesansfil.org/ for a list of such hot spots.

of 3-D models), whereas Nexus via AWML and AWQL emphasize models of the world amenable to querying. However, Nexus applications can be endowed with 2-D maps or 3-D visualizations.

9.5 AWARE SPACES: SMART ENVIRONMENTS AND SMART SPACES

A goal of smart environments as mentioned in Chapter 1 is to endow an environment with sensing and computational capabilities. A space or environment (e.g., a meeting room, an office, a house, a hospital or part thereof, a classroom, a laboratory, etc.) is "smart" in the sense that it can be aware of its current occupants or contents and their actions and behavior, and then such awareness is used to support one or more applications. Sensors are typically embedded into the space so that they do not obstruct the user and the user behaves in a natural manner. At the CSIRO,[9] a smart space is defined as follows:

> A smart space is an environment with numerous elements that sense, think, act, communicate, and interact with people in a way that is robust, self-managing, and scaleable.

The CSIRO Smartlands project[10] aims to use sensors to track livestock, detect unusual animal behavior, perform herding, and monitor animal health.

The National Institute of Standards and Technology in the United States describes a smart meeting room project[11] in which "pervasive devices, sensors, and networks provide infrastructure for context-aware smart meeting rooms that sense ongoing human activities and respond to them." Employing 280 microphones and 7 video cameras, the project aims to capture large amounts of data about meetings. Such data can be processed for speech recognition, speaker identification, gesture recognition, and face identification, which can then be used to support applications such as commanding devices and appliances mentioned in Chapter 6. For example, a command such as "computer, bring up my appointment calendar" can be interpreted (the speaker identified and the relevant application and calendar displayed). Objects being pointed to can also be identified to support such commands. The collected data can be annotated (perhaps semiautomatically, e.g., via speaker recognition) and used for archival purposes.

---

[9] http://www.smartspaces.csiro.au/about.htm.

[10] http://www.smartspaces.csiro.au/applic/smart-lands.htm.

[11] http://www.nist.gov/smartspace/.

In the BiD Smart Space,[12] an array of microphones in the ceiling is used for speaker localization and identification. It is noted that a "smart space is not a single application; it is a platform on which a vast number of applications can be built and tested. The canonical example is automatic adjustment of room conditions, such as for lighting, temperature, or stereo volume, based on detection of user identity and ... preferences."

Classroom 2000 (Abowd, 1999) is one of the earliest projects in which automated multimedia (i.e., audio and video, and contents of electronic whiteboards) capture of lectures is carried out using microphones and video cameras embedded in the ceiling. The captured lectures were then made accessible using an online information system.

In the spirit of Classroom 2000, Labscape (Arnstein et al., 2002) is a system to help "capture formal, detailed representations of laboratory procedures as the work is performed." Touch tablet computers are used by laboratory workers containing information about laboratory procedures and plans. The workers can use the system to record progress and access needed information during their laboratory work. In an implementation, to adequately cover the work areas laboratory workers most often use, five tablet computers were used, distributed throughout the laboratory.

Also interesting is the magic mirror metaphor used in the ALIVE project.[13] A person looks into a screen and sees himself or herself. One camera and vision-based tracking is used to determine the position of the person's head, hand, and foot. The screen mirrors the person's gestures, and additional animated characters can join the person on the screen; the person can interact with these characters. Such a combination of audio and visual input is used for natural human computer interaction in what is called *perceptive spaces* by Wren et al. (1999).

Toward a vision of smart spaces, with the smart house as an example, El-Zabadani et al. (2005) presented a system of smart plugs in which RFID tags (on devices) and readers (at the power outlets) are used to detect when an appliance has been plugged into a power outlet. A map of the power outlets will also reveal roughly the location of the devices. The idea is that new devices added to the room can be detected automatically.

The "intelligent job site" vision proposed by Julien et al. (2005) employs distributed sensors and mobile devices for workers and has interesting features, including:

- A worker can carry a mobile device that interacts with distributed sensors (e.g., read the RFID tag with information about materials, such as a palette of bricks, and record their location).

---

[12] http://www.eecs.berkeley.edu/~davidsun/smartspace/.
[13] http://alive.www.media.mit.edu/projects/alive/.

- Sensors can monitor the site for the amount of hazardous materials and warn workers if the level goes beyond a specified threshold.
- Stress sensors in walls or floors can be used for structural health monitoring.
- A crane has a load sensor to check if it has exceeded its carrying capacity.

A networked distributed software infrastructure with components on the crane computer, hazardous material sensors, workers' mobile devices, and other sensors is described in the reference cited.

These *aware spaces* can provide context information useful for their occupants or context-aware artifacts located within them. Sensing (including manual explicit input by individuals) in an aware space can be viewed as bridging the physical world, which is the space, and the virtual (computational) world, when the sensor information is used to update a computational model mirroring the space or some aspect of it. Such models can then be used to support multiple context-aware applications.

9.6 MIRROR WORLDS: CONTEXT AND ONTOLOGIES

We have previously discussed the use of context models and ontologies in Chapter 2. Ontologies provide the vocabulary of concepts for describing context in context-aware applications. Ontologies can also be used to integrate different context models based on the same ontology. Where different context models are constructed according to different ontologies, mappings between ontologies (Kalfoglou and Schorlemmer, 2003) can be used as a basis to integrate these context models.

To construct mirror worlds in a machine-processable form (amenable to querying, for example) requires a well-defined vocabulary of concepts and relationships among the concepts. In this sense, ontologies are important. An ontology is "a specification of a conceptualization" (Gruber, 1993) of some aspect of the world. As an area of philosophy, ontology has one key question: What are the fundamental categories of being? From this question, one can note immediately the relevance of such an ontology to building mirror worlds. In building a mirror world, a key question is: What needs to go into the mirror world model? A guiding vocabulary of objects provides a good starting point.

Nexus' AWML provides an example of concepts that can be used to model the world, adequate for many applications, which one could view as an ontology. The SOUPA ontology[14] provides concepts for describing context in context-aware applications, including space, action, time,

---

[14] http://pervasive.semanticweb.org/soupa-2004-06.html.

region, location, policy, BDI, agent, schedule, person, and device. Another ontology is CONON (Wang et al., 2004), which includes a general upper ontology containing concepts such as location, person, activity, and computational entities (e.g., service, application, and agent) and specific ontologies for applications (e.g., a home domain). What is interesting is that often such ontologies contain concepts that directly correspond to the physical world yet have existence only in the virtual world.

Ontologies can be used to describe virtual worlds in terms that an end user can understand. For example, VRML (Virtual Reality Markup Language) is used to describe 3-D virtual worlds, but it may be difficult for non-VRNML programmers to create such 3-D worlds. There has been work such as by Bille et al. (2004) that aims to use high-level concepts (described in an ontology) to specify such virtual worlds; the specification is then used to generate the VRML description. For example, in a domain about bowling, the concept "bowling ball" can be mapped to a sphere in VRML. Although such work is still in progress, they illustrate that ontology-based descriptions of 3-D worlds can be used to describe mirror worlds.

9.7 SUMMARY

This chapter has reviewed work related to the notion of mirror worlds, which are virtual counterparts of the real world and parts of physical reality. We have noted similarities across the work in digital cities, projects such as Nexus, and virtual worlds, smart spaces, and environments. We have also pointed out the role of ontologies in constructing not only context models but also mirror worlds. Future work in the area remains, from deployment of model applications of such mirror worlds to solving difficult problems related to the construction and maintenance of large-scale mirror worlds.

REFERENCES

Abowd, G.D., Classroom 2000: an experiment with the instrumentation of a living educational environment. *IBM Systems Journal* 38(4), 508–530, 1999.

Arnstein, L., Borriello, G., Consolvo, S., Hung, C., and Su, J., Labscape: a smart environment for the cell biology laboratory, *IEEE Pervasive Computing* 1(3), 13–21, July–September 2002, IEEE Computer Society.

Bauer, M. and Rothermel, K., How to observe real-world events through a distributed world model, *Proceedings of the 10th International Conference on Parallel and Distributed Systems (ICPADS'04)*, IEEE Computer Society Press.

Benford, S., Bowers, J., Fahlen, L, E., and Greenhalgh, C., Managing mutual awareness in collaborative virtual environments, *Proceedings of VRST'94*, Singapore, 1994, ACM Press, pp. 223–236.

Bille, W., Pellens, B., Kleinermann, F., and De Troyer, O., Intelligent modelling of virtual worlds using domain ontologies, *Proceedings of the Workshop of Intelligent Computing (WIC), held in Conjunction with the MICAI 2004 Conference*, Mexico City, Mexico, 2004, pp. 272–279.

El-Zabadani, H., Helal, A., Abudlrazak, B., and Jansen, E., Self-sensing spaces: smart plugs for smart environments, *Proceedings of the 3rd International Conference on Smart Homes and Health Telematic (ICOST)*, Sherbrooke, Québec, Canada, July 2005.

Ferscha, A., Hechinger, M., Mayrhofer, R., Rocha, D.S., Franz, M., and Oberhauser, R., Digital aura, *Proceedings of the 2nd International Conference on Pervasive Computing*, Vienna, 2004, pp. 405–410.

Gelernter, D., *Mirror Worlds: or the Day Software Puts the Universe in a Shoebox — How It Will Happen and What It Will Mean*, Oxford University Press, U.K., 1993.

Greenhalgh, C. and Benford, S., Massive: a collaborative virtual environment for teleconferencing, *ACM Transactions on Computer-Human Interaction* 2(3), 239–261, 1995.

Gruber, T.R., A translation approach to portable ontologies, *Knowledge Acquisition*, 5(2), 199–220, 1993, available at http://ksl-web.stanford.edu/KSL_Abstracts/KSL-92-71.html.

Holzhauer, D., Creating a Mirror World for Wireless Sensor Networks, available at http://www.afrlhorizons.com/Briefs/Apr04/IF0316.htm [accessed: November 2005].

Ishida, T., Digital city Kyoto: social information infrastructure for everyday life, *Communications of the ACM* 45(7), 76–81, 2002.

Julien, C., Hammer, J., and O'Brien, W.J., A dynamic programming framework for pervasive computing environments, *Proceedings of the Workshop on Building Software for Pervasive Computing at OOPSLA'05*, October 2005, available at http://www.ece.utexas.edu/~julien/pubs/pervasive05.pdf.

Kalfoglou, Y. and Schorlemmer, M., Ontology mapping: the state of the art, *The Knowledge Engineering Review* 18(1), 1–31, 2003.

Lehmann, O., Bauer, M., Becker, C., and Nicklas, D., From home to world — supporting context-aware applications through world models, *Proceedings of the 2nd IEEE Annual Conference on Pervasive Computing and Communications (PERCOM'04)*, 2004, IEEE Computer Society Press.

Nicklas, D. and Mitschang, B., On building location aware applications using an open platform based on the NEXUS augmented world model, *Software and System Modelling* 3(4), 303–313, 2004.

Rodden, T., Populating the application: a model of awareness for cooperative applications, *Proceedings of the Conference on Computer Supported Cooperative Work*, Cambridge, 1996, ACM Press, pp. 87–96.

Wang, X.H., Zhange, D.Q., Gu, T., and Pung, H.K., Ontology based context modeling and reasoning using OWL, *Proceedings of the 2nd IEEE Annual Conference on Pervasive Computing and Communications Workshops*, 2004, IEEE Computer Society Press, pp. 18–22.

Wren, C.R., Basu, S., Sparacino, F., and Pentland, A., Combining audio and video in perceptive spaces, *Proceedings of the 1st International Workshop on Managing Interactions in Smart Environments*, Dublin, Ireland, 1999.

10

CONSTRUCTING CONTEXT-AWARE PERVASIVE SYSTEMS: DECLARATIVE APPROACHES AND DESIGN PATTERNS

Chapter 2 has provided an abstract architecture for context-aware pervasive systems. There have been many ad hoc approaches for constructing specific context-aware systems, and perhaps this has been adequate for the particular application at hand. This chapter explores a general view of such systems — attempting to capture key abstractions common in different context-aware applications.

A key abstraction we consider in this chapter is based on the following question: How does one describe and represent the *situations* that such a system should recognize? For instance, if I were building a context-aware phone, I would like the phone to behave appropriately in certain situations; the phone could somehow detect a situation via some combination of sensors and then switch itself to an appropriate mode (e.g., see that I am in a meeting and put itself to silent mode). Taking a knowledge engineering approach, one could encode knowledge about how to recognize a set of typical situations (or situation types) that the phone can be in, and then rules can be written for appropriate actions in those situations.

Knowledge to recognize a situation typically includes the combination of required sensors and how these sensors are used, appropriate processing and reasoning with sensory and context information, and additional knowledge to interpret and reason about processed information. There is a need for some formalism to represent such knowledge about situations; we are in effect labeling a collection of sensor readings with an interpretation that they represent some situation. In this chapter, we present such an approach to recognizing and reasoning with situations from the perspective of knowledge engineering.

Knowledge engineering has a long history in computing in the context of expert systems, a breed of systems representing a major contribution from the field of Artificial Intelligence. In such systems, knowledge represented in some formal (typically, logic-based) language is systematically encoded into an expert system shell by, or with the help of, a domain expert. With situations, one can rely on common sense to provide knowledge about how they might be recognized, unless the situations are very much specific to an application domain. We (as domain experts) create explicit representations of situations and reason with them. Because situations are recognized via values for context attributes acquired via sensors, the representation given later in this chapter describes situations by relating context attributes and sensors.

Given subsequently is a representation formalism for situations (in the next section) first put forward by Loke (2005), which is based on the principles of logic programming or, more specifically, the language Prolog (Sterling, 1994). We assume a basic knowledge of Prolog in what is to follow, but a reader not acquainted with Prolog can still understand the gist of the exposition, given that Prolog programs are generally readable, as is typical of declarative languages. As an example, we then examine the case study of representing the situation of a meeting, and explore underlying issues. Here, we also point out that the same situation can be represented in many different ways, just as there are many different ways to tell if a particular situation is occurring (or if an entity is currently in a given situation).

Then, we also point out another use of logic programming to reason with sensory information and context. Thereafter, we move from representing situations in a formal language toward broader software engineering concerns, arriving at a preliminary sketch of design patterns for context-aware applications.

10.1 REPRESENTING SITUATIONS

Reiterated here is the operational (and arguably broader) definition of context from Dey (2001): "Context is any information that can be used to characterize the situation of an entity. An entity is a person, place, or object that is considered relevant to the interaction between a user and an application, including the user and applications themselves." Also, studies in such disciplines as situation theory (Barwise and Perry, 1983) and situation calculus consider the primacy of situation abstraction and note that humans can individuate a situation. Situation abstraction allows one to effectively "carve the world up" into manageable pieces, which a collection of sensors of a system might recognize and respond to. It might also be possible to combine such pieces to construct complex representations of situations.

This way of representing situations here takes into account the structure of a context-aware system as comprising sensors at one level and inference procedures to reason with context and situations at another level, following the abstract architecture in Chapter 2. We use a broad definition of sensor as mentioned in Chapter 2, which is taken to mean not only temperature, heat, or motion sensors but any device or mechanism that is used to provide contextual information.

We also consider how to manipulate situations as first-class entities, and how to reason with our representation of situations within a logic programming language. We clearly differentiate between sensor readings, context, and situation in this model.

10.1.1 The Situation Program

Let each sensor be represented by a *sensor predicate* of the form

```
<sensor_id>*(<inputs>, <output_readings>).
```

The output from a sensor is represented by a variable and inputs to sensors by parameters. Then, a *situation program* is defined as a collection of rules (or a logic program), each rule of the form

$$A \ \text{-e->}\ G$$

where "–e->" denotes "is a possible explanation for" (i.e., an abductive reading), and G is given (in EBNF form) by

$$G ::= A \mid S \mid (G, G) \mid S^*{>}E$$

A is an atomic goal formula (an ordinary Prolog-style term), S is a sensor predicate, "," denotes conjunction, S is a situation identifier, and E is an entity (e.g., user, device, or software agent) identifier. We call the operator "in situation" denoted by "*>". A goal of the form $S^*{>}E$, read as a query "E in situation S?," is a meta-level goal, which succeeds if the contextual information about E is provable from S as we describe later. Because S represents clauses (facts and rules that would hold) about the situation, the intuition of this operator is that *E* is in the real-world situation represented by *S* if the contextual information about *E* holds in *S*. There is one distinguished rule (which we call the *situation rule*) whose premise is a predicate naming the situation and, optionally, has a parameter denoting the entity. A rule such as the following

```
fire_in_room(L) -e->
    smoke_detector*(L,positive),
    temperature*(L,R), R> 70.
```

specifies that a fire in room L is a possible explanation for an observed positive reading on the smoke detector in room L, and the temperature reading to be above 70°C. Because the predicate `smoke_detector*/2` is a sensor predicate, it obtains its values from querying the sensor (i.e., the smoke detector). So does the `temperature*/2` predicate.

Predicates of the form `S*>E` allow more complex reasoning to be specified, where a situation occurring might not only explain sensor readings obtained at that time, but also clarify why some other situations have been inferred as occurring (through other abductive rules). For example, the following rule says that E being in a situation `sleeping/1` is a possible explanation for E being in situation `not_answering_calls/1` and E being in situation `motionless/1`:

```
sleeping(E) -e->
    not_answering_calls*>E,
    motionless*>E.
```

Of course, there could be other reasons why E is not answering calls and E is motionless, such as

```
dead(E) -e->
    not_answering_calls*>E,
    motionless*>E.
```

Such is the nature of abductive reasoning, in that more than one possible explanation might exist for given observations. Several possibilities might be presented to the user as equally possible answers, or the system might attempt to choose the best explanation based on some other strategy; for example, try to prove that E cannot be dead. We do not dwell further on strategies for choosing explanations here.

The rules of a situation program permit natural representation of a situation; i.e., a situation occurring helps to explain certain observations and constraints that exist, including the existence of some other situation.

As another example, we can define a in_meeting_now situation as follows. The sensor predicates are location*(E,L), which returns the location of an entity E in variable L; diary*(E, Event, entry (StartTime, Duration)), which returns diary entries for entity E for a matching event; people_in_room*(L,N), which returns the number of people at a location; and current_time*(T), which takes no inputs and returns the current time in a variable. The constraints the situation imposes on such sensors' readings can then be modeled by the following logic program:

```
situation program meeting1:
in_meeting_now(E)  -e->
    with_someone_now(E),
    has_entry_for_meeting_in_diary(E).
with_someone_now(E)  -e->
    location*(E,L),
    people_in_room*(L,N),  N > 1.
has_entry_for_meeting_in_diary(E)  -e->
    current_time*(T1),
    diary*(E,'meeting',entry(StartTime,Duration)),
    within_interval(T1, StartTime, Duration).
```

The program is viewed as a constraint in the sense that if the entity is in that situation, various relationships as specified earlier should hold.

10.1.2 Modularity

As mentioned, syntax of rules just mentioned allows situation programs that refer to other situation programs. The same program might be rewritten as follows:

```
situation program meeting1:
in_meeting_now(E)  -e->
    with_someone_now*>E,
    has_entry_for_meeting_in_diary*>E.
```

where with_someone_now is a situation with its own situation program containing the rule

```
with_someone_now(E)  -e->
    location*(E,L),
    people_in_room*(L,N),  N > 1.
```

and similarly, the situation `has_entry_for_meeting_in_diary`.

The ability to split rules into separate situation programs leads to the advantage of modularity, which encourages reuse. Hence, the operator "*>" might simply be viewed as a mechanism to refer to another situation program, i.e., `with_someone_now*>E` is different from `with_someone_now(E)` only in that the former refers to another situation program, whereas the latter should be inferred using rules in the same situation program.

Because we represent situations as explanations for observations, the procedure for evaluating the in-situation goal is by forward-chaining over rules in situation programs, as described by Loke (2005). Evaluating such goals will involve querying the sensors and then checking, given the sensor readings obtained, if the relationships specified in a situation program are satisfied.

10.2 FIVE OTHER WAYS TO REPRESENT A MEETING

We have shown one way to represent the situation of a meeting occurring. In general, if we use a different set of sensors, we can define a different situation program for a meeting. Five other possibilities are given as follows, with different assumptions about the situations, the sensors used, and what sensory information can be obtained (e.g., what objects or people are being tracked):

1. *Colocation of filled coffee cups in a room*, as inspired by Gellersen et al. (2002): We assume a database of coffee cups whose location is tracked by a positioning technology, retrieved when given a name, using the predicate `has_coffee_cup/2`, and a database of employees and their colleagues, retrieved when given a name, using the predicate `has_colleagues/2`. We have the following rule which says that E is in a meeting with at least one other colleague explains why E's coffee cup and at least one of E's colleagues' coffee cups are colocated in the same room, and are warm (above 50°C):

   ```
   situation program meeting2:
   in_meeting_now(E)  -e->
       has_colleagues(E, Fs),
   ```

```
member(F, Fs),
has_coffee_cup(F, CF),
has_coffee_cup(E, CE),
location*(CE, Room),
location*(CF, Room),
temperature*(CF, TCF),
temperature*(CE, TCE),
TCF > 50, TCE > 50.
```

The rule works on real-world assumptions about coffee usage at meetings. Another view is that this rule defines a specific kind of meeting where people use coffee.

2. *Weight sensors on the floor:* We assume a weighing machine on the floor of a room, which gives the total weight of objects, including people, on it. A rule such as the following states that a meeting is occurring in the room E is located in explains why the weighing machine of the room would have a reading above some threshold.

```
situation program meeting3:
in_meeting_now(E)  -e->
    location*(E, Room),
    floor_weight_machine*(Room, W),
    W > 200.
```

3. *Devices in the room:* Similar to Ranganathan and Campbell (2003), we assume that the fact that a meeting is going on explains why lights will be on in the room, PowerPoint is running on the PC in the room, and the projector is working. The following rule captures this idea about the room that a person, E, is currently in:

```
situation program meeting4:
in_meeting_now(E)  -e->
    location*(E, Room),
    projector*(Room, switched_on),
    room_light*(Room,switched_on),
    pc_software_applications*(Room,
        powerpoint, running).
```

We have assumed predicates that will return the status of devices and PC applications.

4. *Sounds and noises:* We assume the presence of microphones in the room which measure noise levels. Our assumption is that compared to other times when noise levels are generally low, the

noise level when a meeting is going on will be significantly higher. Noise levels can be measured by a noise dosimeter[1] worn by a person or a sound level meter situated in a meeting room of interest. We can perhaps do better by matching sounds with voice patterns of speakers, with the person of interest, E, to see if this person is present in the meeting, perhaps even analyzing the person's speech by noting keywords. Assuming a dosimeter worn by a user (say, by E), the following rule stating that the person is in a meeting explains why the noise level as detected by a dosimeter worn on the person is above a certain level averaged over a period of time (say measured over 5 min since the time of the start of the query to determine if E is in a meeting) and why the average sound readings from a sound meter in the room is above a threshold for a similar period of time (such a threshold serves to distinguish the meeting situation in the room from other times when it is typically quiet in the room) — some calibration of the meters are required to determine the thresholds:

```
situation program meeting5:

in_meeting_now(E, PersonalMeetingNoiseThreshold)
-e->

    dosimeter*(E, 5, AvgNoiseLevel),
        % average noise level returned
        % after 5 minutes of measurement
    AvgNoiseLevel > PersonalMeetingNoiseThreshold,
    location*(E,Room), % check noise
        % level in the room the person is in
    meeting_room(Room, RoomMeetingSoundThreshold),
        % ensure that the room is a meeting room and
        % retrieve the threshold
    sound_meter*(Room, 5, AvgSoundLevel),
        % average sound level measured over 5
        minutes
    AvgSoundLevel > RoomMeetingSoundThreshold.
```

5. *Use cameras:* Cameras can be used to detect the presence of people in a meeting room using a technique called background substraction.[2] This technique can be used to "watch" meeting rooms for activity.

---

[1] See, for example, http://www.quest-technologies.com/Noise/index.htm.
[2] See http://www.flong.com/writings/texts/essay_cvad.html.

```
situation program meeting6:
in_meeting_now(E)  -e->
    location*(E,Room),
    people_present*(Room).
    % this is done by a camera watching the room
```

10.2.1 Observations

We make the following observations, based on the foregoing examples:

- *Multiple representations*: We note that the same situation can be represented in multiple ways and that we can combine multiple situation programs in modeling a given situation.
- *Abductive view*: Each situation program captures some aspect of the situation of a meeting occurring, but each might be viewed as not totally conclusive; i.e., if all the relationships specified in a particular situation program hold, one could guess that a meeting is occurring to a high degree of certainty, though not with absolute certainty. Hence, we view these situation programs abductively: a meeting occurring is a possible explanation that the relationships specified in a particular situation program are being observed as occurring.
- *Modular representations*: Because each situation program contains a set of relations, which should hold given that a situation occurs, one can devise different situation programs (containing different sets of relationships) for modeling situations. As we have seen with the in-situation operator, one situation program can refer to others. Also, one can build sophisticated representations of (more complex) situations in terms of an existing repertoire of situations.
- *Programming with situations*: The situation programs can be embedded into logic programs using metaprogramming (i.e., manipulating situation programs as first-class entities within programs), and reasoned about within a declarative framework. An example of this technique is provided in the following section.
- *Design patterns*: From the software engineering perspective, there is indication from the foregoing examples that design patterns can be developed for recognizing situations. Taking a situation program, one can elaborate on it to spell out particular sensors to be used and what reasoning techniques might be employed to validate or, as is the case, to invalidate (with respect to the real world) each of the mentioned relationships. Based on the same template of a situation program, different reasoning techniques might be employed to recognize different relationships in different applica-

tions. For the same situation program, different sensors and underlying technology might also be employed according to what is available; i.e., situation programs can represent situations at a reasonable level of abstraction, decoupled from underlying technologies. We provide a sketch of such a design pattern later in this chapter.

The preceding discussion advocates a high-level explicit representation of situations for the purposes of context-aware pervasive computing, approached from the traditional knowledge engineering perspective. The proposed situation program formalism should be viewed as illustrative, not prescriptive; other approaches and formalisms can be employed (for example, we review two other declarative approaches later) but, we contend, should retain the spirit of this approach. The benefits of such an approach include enabling abstraction from underlying sensor technologies, modularity and reuse of representations, and metaprogramming style manipulation of situation representations.

10.3 METAPROGRAMMING WITH SITUATION PROGRAMS: EXAMPLES

The in-situation operator "*>" can be embedded into Prolog programs as a distinguished predicate, evaluated to infer whether an entity E is in a situation defined by a situation program S. We can then write rules involving this operator. We call Prolog extended with this use of the operator *metaprograms* because the rule references other (situation) programs.

The following example is a rule that states a meeting is currently not possible among a list of individuals when any one of them is at home.

```
meeting_not_currently_possible(Es):-
    member(E,Es),
    at_home*>E.
```

Prolog backtracking search applied with the rule will go through every member of the list Es and returns true for "the goal meeting not possible" if any one of the members of Es is found to be at home. Note that normal Prolog evaluation will be employed for evaluating such rules, except when evaluating the in-situation goals (e.g., at_home*>E), where a different evaluation strategy (as noted earlier) will be employed for these goals.

Another example rule here specifies if an individual E is in one situation from a given list:

```
one_situation(Ss, E):-
    member(S, Ss),
    S*>E.
```

This simple rule illustrates metaprogramming, with situation programs Ss manipulated as first-class entities.

Mapping of situations to actions can be readily encoded. For example, the following rules describe different actions to take when in different situations in a context-aware messaging application:

```
action(storeMessage, E):-
    in_meeting*>E.
action(forwardMessage, E):-
    at_home*>E.
action(playMessage, E):-
    available*>E.
```

A query such as ?- action(WhatToDo, john) will instantiate the variable WhatToDo with an action suitable for john's situation. Default actions can also be similarly defined.

Complex situations can be defined in terms of existing situation programs. For example, the following rule defines an entity E as not interruptible if E is in any one of four situations (not exhaustive):

```
not_interruptible(E):-
        bathing*>E
    ; in_meeting*>E
    ; sleeping*>E
    ; driving*>E.
```

Note that this presents another way of inferring situations, not abductively as explanations for observations, but as an abstract situation inferred, given that one has inferred other situations (from say, observed sensor readings). The full expressiveness of a Turing complete programming language such as Prolog can be brought into use for inferring situations.

10.4 ANOTHER DECLARATIVE APPROACH

This section describes another application of the declarative programming paradigm to the problem of representing and reasoning with sensory information and context.

This approach is termed *semantic streams* (Whitehouse et al., 2006) and aims to provide a declarative, logic-based framework for querying sensor networks. Although this work is still ongoing at the time of writing of the book, we provide a sketch of the idea as follows.

The idea is to view sensors as services providing sensory data and to "wrap up" sensors as logic predicates describing sensor data streams. For example, a Vehicle Detector in a parking deck can be described as a service (viewed as a rule) that uses a magnetometer sensor (preconditions of the rule) to detect vehicles and creates an event stream providing the time and location of detected vehicles (postconditions of the rule), using the following predicate:

```
service(magVehicleDetectionService,
    needs(
        sensor(magnetometer, R)
    ),
    creates(
        stream(X),
        isa(X, vehicle),
        property(X, T, time),
        property(X, R, region)
    )
).
```

Predicates of the form

```
sensor(<sensor type>, <region>)
```

define the type and location of each sensor. For example, for the magnetometer used in this case, we have a fact of the form

```
sensor(magnetometer, [[60,0,0], [70,10,10]]).
```

which declares a magnetometer sensor as covering a region corresponding to a 3-D cube defined by a pair of [x,y,z] coordinates, i.e., vehicles within this region are detected by the sensor. Detection of vehicles leads to instantiation of the events stream in variable X.

Services can be composed by matching the preconditions of a service with the postconditions of another service. For example, the following service takes an event stream from the preceding service and produces a histogram:

```
service(histogramService,
    needs(
        streams(X),
        isa(X, vehicle),
        property(X, T, time)
    ),
    creates(
        stream(Y),
        isa(Y, histogram)
    )
).
```

Note that the preconditions of the `histogramService` matches the postconditions of the `magVehicleDetectionService`.

Consider a query to return an event stream of detected vehicles, such as

```
?- stream(X), isa(X, histogram).
```

This query will be evaluated by a process similar to backward-chaining, creating a chain of services in which the postconditions of a service matches the preconditions of another service. For example, the foregoing query will match the postconditions of the `histogramService`, which leads to a subgoal that comprises the preconditions of the `histogram-Service`, which, in turn, as we have seen, will match the postconditions of the `magVehicleDetectionService`. Instantiating and executing the `magVehicleDetectionService` leads to low-level queries on the magnetometer sensors.

The semantic streams approach differs from situation programs, in that it does not seek to represent the notion of situations, but through the declarative programs interprets sensory data at a high level of abstraction. For example, the detection or presence of vehicles is represented declaratively and can be combined into a knowledge base which maps collected context information (such as the presence of vehicle) to some high-level situation. Finally, we note the modularity achieved in the semantic streams approach using the service abstraction.

10.5 TOWARD DESIGN PATTERNS FOR CONTEXT-AWARE APPLICATIONS: SITUATION PATTERNS

These examples of situation programs capture knowledge about recognizing particular situations, either abductively from observed sensor readings and relationships between such readings or deductively given that certain situations are assumed to have been recognized. A situation program describes a way in which a type of situation can be recognized. It is possible to apply the same situation program in different places to recognize the same type of situation.

For example, if we consider the situation program to recognize meetings via the colocation of coffee cups as given earlier, then that situation program can be applied to one's own office in which there are sensors to track the location of the relevant coffee cups. The same situation program might be applied with a different set of coffee cups in someone else's office. Hence, it is in this sense that a situation program can be used to recognize a type of situation (i.e., meetings) as opposed to merely recognizing a particular meeting happening at a certain place and at a certain point in time. But the particular application of a situation program depends on the lower-level sensors (in this case, the actual set of coffee cups and the associated tracking technology used), and relies on the interface made between what is referenced in the situation program and the actual sensors available. Hence, a situation program can be considered a high-level template or pattern for recognizing a type of situations.

Design patterns are a popular design artifact for object-oriented programming for capturing reusable programming idioms and techniques, applicable to different software systems (Gamma et al., 1995). In essence, a design pattern records experience in designing systems, so that others who want to build something similar need not start from scratch.

According to Christopher Alexander,[3]

> Each pattern describes a problem which occurs over and over again in our environment and then describes the core of the solution to that problem in such a way that you can use this solution a million times over, without ever doing it the same way twice.

A design pattern is more than a programming template but often contains documentation concerning the design. From Gamma et al. (1995), on object-oriented software design patterns:

> Each design pattern systematically names, explains, and evaluates an important and recurring design in object-oriented systems. Our goal is to capture design experience in a form that people can use effectively.

---

[3] http://www.patternlanguage.com/leveltwo/caframe.htm?/leveltwo/../bios/designpatterns.htm.

We sketch here ideas toward developing design patterns for recognizing situations, or what we term *situation patterns*.

Suppose one is designing such a system, one would need to think of what context information is required, the sensors to use to acquire such context and how they should be organized, what kind of reasoning to use for processing the sensor results, and effectiveness of such a setup. Experiences in the design of this system can be captured in a situation pattern. A template might contain the following information, based on object-oriented design patterns:

Pattern name: A name that conveys the idea of the pattern succinctly.

Intent: A short statement that captures the rationale and intent of the pattern, and what problem the pattern intends to solve. What situation or situations does the pattern aim to recognize?

Motivation: What motivated the pattern in the first place, as well as what motivated the method for solving the problem as proposed by the pattern?

Applicability: What are circumstances under which the pattern is useful? Are there specific applications where the pattern is particularly helpful?

Context and sensors: What type of sensors are used and what contextual information are the sensors used to acquire?

Processing and reasoning: How is the data from the sensors processed — what specific data analysis (e.g., data mining or learning) techniques are used? Also, is there reasoning about context to infer if some situation is occurring? What kind of reasoning is used?

Consequences: What is the outcome of using the pattern? The effectiveness of the pattern is reviewed.

Implementation: Discuss the actually implemented case study, highlighting practical considerations when applying the pattern to implement a solution. What are the issues to watch out for during implementation; i.e., what are the practical considerations? What difficulties could be encountered? What are the actual sensors used and how many? What hardware and software are used with the sensors? What is the setup and the physical characteristics of the place where the sensors have been deployed?

Sample code, known uses, and related work and patterns: Sample code from the implementation can be attached here. Also, known uses of the pattern can be recorded, as well as related patterns for recognizing similar situations.

The first example is a pattern which is developed based on the work by Fogarty et al. (2005) on how to recognize when it is appropriate to interrupt someone (say, in the office environment), i.e., to predict human interruptibility.

Pattern name: Simple estimation of human interruptibility.

Intent: The pattern aims to capture a solution for effectively recognizing the situation whereby a human being can be interrupted in a socially appropriate manner.

Motivation: The aim is to enable proactive devices that may interrupt users, such as mobile phones, e-mail, and other messaging applications, but in a socially considerate way. This would then not require the user to disable the devices manually.

Applicability: The pattern focuses on users in everyday office situations. An experimental implementation of the pattern as described here shows that the pattern can provide estimates of human interruptibility as good as or better than estimates provided manually by humans watching audio and video recordings of the environment.

Context and sensors: The sensors used are:

- A microphone in the corner of an office to detect noise level (e.g., to determine if the person is talking), calibrated against background noise.
- Computer clock to return the time of day.
- A sensor to detect when the phone is in use.[4]
- Computer returning activity information for the mouse and keyboard (to determine if the person is busy at the computer).

Readings from such sensors can be combined to estimate human interruptibility.

Processing and reasoning: Data from the sensors are processed to give an estimate of how busy the user currently is. Calibration of the microphone is required to determine background noise. The sensor to detect phone use and computer input activity indicates that the user is doing something and perhaps should not be interrupted. Levels of interruptibility can also be defined according to the sensor readings and the user's activity.

Consequences: The pattern can be used to estimate interruptibility as accurately as human estimators can, using a set of low-cost sensors with simple data processing.

Implementation: Experimentation as documented in Fogarty et al. (2005) was done using simulated sensors for the telephone, keyboard, and mouse.

---

[4] http://www.maxim-ic.com/appnotes.cfm/appnote_number/3521 contains information about a device called a "telephone nanny" which can track such phone usage.

Sample code, known uses, and related work and patterns: Other studies such as Vorburger and Bernstein (2005) have employed other sensor data such as video data besides audio and time of day. Video data can be analyzed for the presence of one or more people in the office.

The preceding pattern example lacks detail in the implementation (e.g., details of the microphone and phone use sensor used), given that it is derived from the study cited, which simulated some of the sensors. Actual deployments of the system with sensors can provide further information under the implementation section of this pattern. Moreover, there has been other work on using cameras to estimate human interruptibility, and such work can be captured in a similar pattern. Other context-aware applications can be similarly captured in a design pattern, such as recognizing a meeting, instructions to print addressed to the nearest printer, or find the nearest ATM.

10.6 SUMMARY

This chapter has reviewed declarative approaches to constructing context-aware applications and has considered design patterns for context-aware applications, involving an explicit representation of know-how in recognizing situations and documentation of such know-how in two different forms, as well as involving different emphases. The idea in both, however, is to illustrate how to capture reusable knowledge to facilitate building this new breed of applications.

ACKNOWLEDGMENT

This chapter contains portions reprinted with permission from "On representing situations for context-aware pervasive computing: six ways to tell if you are in a meeting by Loke." PerCom Workshops 2006, pp. 35–39 ©2006 IEEE.

REFERENCES

Barwise, J. and Perry, J., *Situations and Attitudes*, Cambridge, MA: MIT-Bradford, 1983.

Dey, A.K., Understanding and using context, *Personal and Ubiquitous Computing Journal* 5(1), 5–7, 2001, Springer.

Fogarty, J., Hudson, S.E., Atkeson, C.G., Avrahami, D., Forlizzi, J., Kiesler, S., Lee, J.C., and Yang, J., Predicting human interruptibility with sensors, *ACM Transactions on Computer-Human Interaction (TOCHI)* 12(1), 119–146, March 2005.

Gamma, E., Helm, R., Johnson, R., and Vlissides, J., *Design Patterns: Elements of Reusable Object-Oriented Software*, Addison Wesley Longman, Inc., U.S.A., 1995.

Gellersen, H.-W., Schmidt, A., and Beigl, M., Multi-sensor context-awareness in mobile devices and smart artifacts, *Mobile Networks and Applications (MONET)* 7(5), 341–351, 2002.

Loke, S.W., Representing and reasoning with situations for context-aware pervasive computing: a logic programming perspective, *The Knowledge Engineering Review* 19(3), 213–233, 2005, Cambridge University Press.

Ranganathan, A. and Campbell, R.H., An infrastructure for context-awareness based on first order logic, *Personal and Ubiquitous Computing Journal* 7, 353–364, 2003, Springer.

Sterling, L., *The Art of Prolog*, Cambridge, MA: MIT Press, 1994.

Vorburger, P. and Bernstein, A., Towards an Artificial Receptionist: Anticipating a Persons Phone Behavior, Technical Report, University of Zurich, Department of Informatics, 2005, available at http://www.ifi.unizh.ch/ddis/staff/goehring/btw/files/tech_vorburger_large.pdf.

Whitehouse, K., Zhao, F., and Liu, J., Semantic streams: a framework for composable semantic interpretation of sensor data, *Proceedings of the 3rd European Workshop on Wireless Sensor Networks (EWSN 2006)*, Zurich, Switzerland, 2006, pp. 5–20.

11

A FUTURE WITH AWARE SYSTEMS

Our surroundings resonate not only with living beings but also with nonliving objects that exhibit the behavior of living beings. Aware systems aim at enriching our lives with new experiences too difficult to achieve in the past but possible in this era of the computer. But it is not only experiences but also the new functionalities and conveniences that aware systems can enable which make them an important development today and in time to come. In some ways, awareness might be thought of as a relabeling of various kinds of existing functionality on one hand and as a concept too difficult to fully realize on the other.

This book contends that aware systems are a useful metaphor for practical systems, a philosophy about what can be, and has sought to examine this concept of context awareness as manifested in different areas of computing. Architectural considerations have been given, and examples of software architectures for implementing context awareness in diverse settings have been illustrated. We can observe the commonalities across different application areas and see cross-cutting concerns.

11.1 THE EMERGING FUTURE: TAKING AWARENESS FOR GRANTED

What does a future with aware systems look like? Bring home new furniture and it changes color and pattern like a chameleon to blend beautifully into the living room. Wrestling with computers and appliances, we get error messages that we always understand and can do something about, irrespective of our background knowledge or language. Our watches and clocks always tell the right time, wherever we are. Devices no longer rudely interrupt us but merely amplify our senses and our basic human capabilities. Systems seem to know what is going on, even if we do not. With aware cars and roads, no one dies in road accidents, unlike today. After repeating the contents of previous chapters, the list goes on.

11.2 SCALABILITY AND USABILITY

Two concerns that we will consider are scalability and usability. Another topic of interest would be synergies among context-aware artifacts from effects that one artifact could have on another. For instance, if we have blinds that automatically adjust themselves according to changing sunlight and a preset measure of light within the house with devices that react to changes in light levels, then there could be cascading effects, where the actions of some device change the context for another device, thus triggering another action, which, in turn, changes the context for some other device, triggering further actions. Another scenario is a change in context such that multiple devices respond (perhaps not in a coordinated way), albeit this might not be a problem depending on the nature of the responses.

How should users behave in a world of aware objects and places? Users can have their actions and behaviors observed, interpreted, and responded to by aware objects and places. But users will want ultimately to remain in control — not just in the case of device failures — and might prefer manual systems with full control over automatic behaviors without control, given a need to choose between the two. However, somewhere in between might be ideal, where users are ready to accept automatic behaviors, sacrificing some control for convenience and compliance reasons.

What would favor automation is that the actions taken in response to context triggers are reversible or that the actions taken do not have severe consequences. Rules which map context triggers to actions can be labeled with such information about consequences. A proposal is to have rules of the following forms:

```
IF Uncertainty(Context) < U
and Severity(Action) < S
THEN DO Action
```

which states that if uncertainty in ascertaining a context (e.g., a location) is less than a given threshold and the severity of action (e.g., send an advertisement) based on some predefined scale is less than a given threshold, then the action is performed (without consulting the user), or

```
IF Uncertainty(Context) > U' and Severity(Action)
> S'
THEN DO Ask-User
```

which states that if the uncertainty in ascertaining the context is more than a threshold and the severity of action is more than a given threshold, then the action is to ask the user first rather than performing the action immediately. Such control rules can map context to actions in some situations, consult the user sometimes, and perhaps not take any action in others. Hence, sets of possible actions, rather than merely binary decision rules, can be considered involving the user to different extents (and, correspondingly, automation to different extents). User interface and usability issues for context-aware computing, of course, would require application-specific solutions and actual user testing.

11.3 FINAL WORDS

The reader will, perhaps, not find it too difficult to think of software or everyday artifacts not mentioned in this book that can and should be aware but are not. After all that has been said, readers will have the opportunity to implement the designs in this book for their respective applications, modify and adapt the designs to suit their own purposes, or invent and design new context-aware objects to distinguish themselves in the market. Innovations or improvements can perhaps happen when one starts to ask, "What if X is an aware system?" — whatever X may be — and ask this question often enough. A context-aware X can also be context-aware in different ways, limited only by the imagination. Today's aware X might be a far cry from the future aware X. We are expecting a future full of aware systems, and once enough of them pervade our lives, we might actually become unaware of such aware systems. Mothers in the not-so-distant future can look forward to the aware bicycle and an enthusiastic cry from their child, "Look ma, no hands, and no hands needed!"

INDEX